Smart Materials and New Technologies

For the architecture and design professions

D. Michelle Addington
Daniel L. Schodek
Harvard University

AMSTERDAM • BOSTON • HEIDELBERG • LONDON • NEW YORK • OXFORD
PARIS • SAN DIEGO • SAN FRANCISCO • SINGAPORE • SYDNEY • TOKYO
Architectural Press is an imprint of Elsevier

Architectural Press
An imprint of Elsevier
Linacre House, Jordan Hill, Oxford OX2 8DP
30 Corporate Drive, Burlington, MA 01803

First published 2005

British Library Cataloguing in Publication Data
A catalogue record for this book is available from the British Library

Library of Congress Cataloguing in Publication Data
A catalogue record for this book is available from the Library of Congress

ISBN 0 7506 6225 5

Contents

Preface

Ten years ago, when we first began treading in the murky waters of ''smart'' materials and micro-systems, we had little information to guide us. Although there had already been rapid expansion in these technologies in the science and engineering fields, particularly in regard to sensor development, their entry into the design arena was, at best, idiosyncratic. We found many novelty items and toys – mugs that changed color when hot coffee was poured inside, and rubber dinosaurs whose heads bobbed when connected to a battery – and we noted that many designers were beginning to incorporate the language of smart materials, albeit not the technologies themselves. There were proposals for buildings to be entirely sheathed with ''smart'' gel, or for ''smart'' rooms that would deform individually for each occupant according to their specific physiological and psychological needs. Precisely *how* this would happen remained mysterious, and it was often presumed that the magical abilities attributed to the smart designs were simply technicalities that someone else – an engineer perhaps – would figure out.

These proposals troubled us from two aspects. The first was clearly that designers were considering these very new and sophisticated materials and technologies to fit right into their normative practice, making design simpler as the manifestation of intentions could shift from the responsibility of the designer to the material itself. One would no longer have to carefully and tediously design wall articulation to create a particular visual effect, as the material would be capable of creating any effect, one only had to *name* it. In addition to this abdication of responsibility to an as-yet undefined technology, we were also concerned with the lack of interest in the actual *behavior* of the technology. By framing these technologies from within the design practice, architects and designers were missing the opportunity to exploit unprecedented properties and behaviors that should have been leading to radically different approaches for design rather than only to the manifestation of designs constrained by the hegemony of existing practice.

When we looked at the other end of the spectrum to examine what scientists and engineers were doing, however, we encountered equally problematic responses. Much of the early development had been geared toward miniaturization and/or simplification of existing technologies – using instantaneous labs on a chip to reduce the time of the unwieldy chromatography process; replacing complex mechanical valves with seamless shape memory actuators. As manufacturing processes were adapted to these specialized materials, and advances in imaging allowed fabrication at the nano scale level, the development shifted from problem solving to "technology push." Countless new materials and technologies emerged, many looking for a home, and a potential application.

We were confronted with trying to fit round pegs – highly specific technologies – into square holes – incredibly vague architectural aspirations. Neither end seemed appropriate. We did not have the kind of problems that a new technology could easily step in to solve, nor did we have any idea about just what kind of potential could be wrung from the behaviors of these technologies. We needed to bridge the very large gap between the owners of the relevant knowledge and the inventors of the potential applications.

This transfer of knowledge has not been easy. Scientific and engineering information typically enters the design realm already "dumbed down." Architects and designers don't need to know how something works, they just need to know the pragmatics – how big is it, what does it look like? This approach, unfortunately, keeps the design professions at arm's length, preventing not only the full exploitation of these technologies, but also denying a coherent vision of the future to help direct development in the science and engineering disciplines. Over the last ten years, we have struggled in our own research, and in our classes, to find the fluid medium between knowledge and application, so that both are served. This book represents the culmination of that decade of investigation and experimentation.

Our primary intention for the book's content was the development of a coherent structure and language to facilitate knowledge transfer at its highest level. There are certain phenomena and physical properties that must be fully understood in order to design a behavior. Fundamental for architects and designers is the understanding that we cannot frame these technologies within our own practice, we must instead inflect their deployment based on their inherent characteristics. For example, as evidenced by the continuing desire of architects to produce smart facades, we have a tendency to ask these technologies to act at our normative scale – the scale of a building. Most of these technologies, however, perform at the molecular and micro-scales. How differently might we think and design if we engaged these scale differences rather than ignoring them?

Clearly, the knowledge about these materials and technologies within the science and engineering realms is so vast that any given engineer will have a different knowledge set than another, even in the same area of specialty. What knowledge, then, should we bring across the divide to the designers? We identified some fundamental laws of physics and principles of materials science that we felt could serve as the building blocks to allow the derivation of behaviors most relevant to the design professions. Several different materials, components and assemblies were then chosen and described to illustrate how these building blocks could be applied to help understand and ultimately exploit each example's characteristics. We fully expect that the specific materials and technologies referred to in this book will soon become obsolete, but we strongly believe that the theoretical structure developed herein will transcend the specifics and be applicable to each new material that we may confront in the future.

Michelle Addington
Cambridge, Massachusetts

Acknowledgments

We are grateful to the many students over the last decade who have willingly experimented with unfamiliar materials and technologies in our courses as we explored the untapped possibilities inherent in thinking about architecture as a network of transient environments. A number of these students have directly supported the development of this book; in particular, our teaching assistants and fellows: John An, Nico Kienzl, Adriana Lira, Linda Kleinschmidt, and Andrew Simpson. Nico, as our first doctoral student in the area, was instrumental in helping us transition to more direct hands-on workshops for the students, and John, our most recent doctoral student in the area, spearheaded a spin-off course that uses simulation techniques. We would also like to thank the two chair-persons of the architecture department – Toshiko Mori and Jorge Silvetti – who supported the development of coursework in this area that helped lead to this book. And always, we are fortunate to have excellent faculty colleagues that we invariably rely upon for support, including Marco Steinberg, Martin Bechthold, and Kimo Griggs.

Michelle Addington and Daniel Schodek

1 Materials in architecture and design

Smart planes – intelligent houses – shape memory textiles – micromachines – self-assembling structures – color-changing paint – nanosystems. The vocabulary of the material world has changed dramatically since 1992, when the first 'smart material' emerged commercially in, of all things, snow skis. Defined as 'highly engineered materials that respond intelligently to their environment', smart materials have become the 'go-to' answer for the 21st century's technological needs. NASA is counting on smart materials to spearhead the first major change in aeronautic technology since the development of hypersonic flight, and the US Defense Department envisions smart materials as the linchpin technology behind the 'soldier of the future', who will be equipped with everything from smart tourniquets to chameleon-like clothing. At the other end of the application spectrum, toys as basic as 'Play-Doh' and equipment as ubiquitous as laser printers and automobile airbag controls have already incorporated numerous examples of this technology during the past decade. It is the stuff of our future even as it has already percolated into many aspects of our daily lives.

In the sweeping 'glamorization' of smart materials, we often forget the legacy from which these materials sprouted seemingly so recently and suddenly. Texts from as early as 300 BC were the first to document the 'science' of alchemy.[1] Metallurgy was by then a well-developed technology practiced by the Greeks and Egyptians, but many philosophers were concerned that this empirical practice was not governed by a satisfactory scientific theory. Alchemy emerged as that theory, even though today we routinely think of alchemy as having been practiced by late medieval mystics and charlatans. Throughout most of its lifetime, alchemy was associated with the transmutation of metals, but was also substantially concerned with the ability to change the appearance, in particular the color, of given substances. While we often hear about the quest for gold, there was an equal amount of attention devoted to trying to change the colors of various metals into purple, the color of royalty. Nineteenth-century magic was similarly founded on the desire for something to be other than it is, and one of the most remarkable predecessors to today's color-changing materials was represented by an ingenious assembly known as a 'blow book'. The magician

▲ **Figure 1-1** NASA's vision of a smart plane that will use smart materials to 'morph' in response to changing environmental conditions. (NASA LARC)

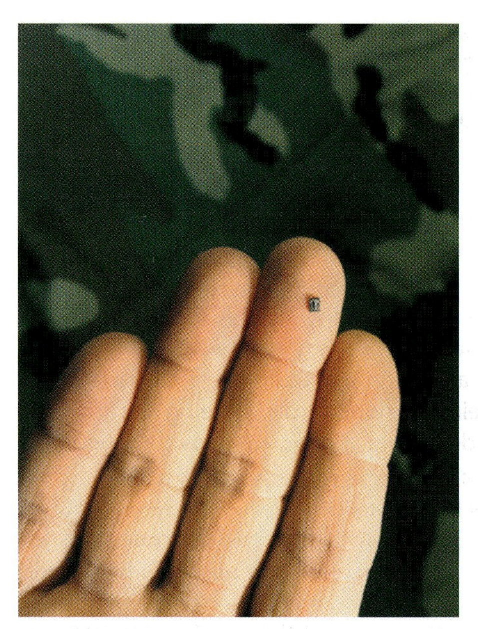

▲ **Figure 1-2** Wireless body temperature sensor will communicate soldier's physical state to a medic's helmet. (Courtesy of ORNL)

would flip through the pages of the book, demonstrating to the audience that all the pages were blank. He would then blow on the pages with his warm breath, and reflip through the book, thrilling the audience with the sudden appearance of images on every page. That the book was composed of pages alternating between image and blank with carefully placed indentions to control which page flipped in relation to the others makes it no less a conceptual twin to the modern 'thermochromic' material.

What, then, distinguishes 'smart materials'? This book sets out to answer that question in the next eight chapters and, furthermore, to lay the groundwork for the assimilation and exploitation of this technological advancement within the design professions. Unlike science-driven professions in which technologies are constantly in flux, many of the design professions, and particularly architecture, have seen relatively little technological and material change since the 19th century. Automobiles are substantially unchanged from their forebear a century ago, and we still use the building framing systems developed during the Industrial Revolution. In our forthcoming exploration of smart materials and new technologies we must be ever-mindful of the unique challenges presented by our field, and cognizant of the fundamental roots of the barriers to implementation. Architecture heightens the issues brought about by the adoption of new technologies, for in contrast to many other fields in which the material choice 'serves' the problem at hand, materials and architecture have been inextricably linked throughout their history.

1.1 Materials and architecture

The relationship between architecture and materials had been fairly straightforward until the Industrial Revolution. Materials were chosen either pragmatically – for their utility and availability – or they were chosen formally – for their appearance and ornamental qualities. Locally available stone formed foundations and walls, and high-quality marbles often appeared as thin veneers covering the rough construction. Decisions about building and architecture determined the material choice, and as such, we can consider the pre-19th century use of materials in design to have been subordinate to issues in function and form. Furthermore, materials were not standardized, so builders and architects were forced to rely on an extrinsic understanding of their properties and performance. In essence, knowledge of materials was gained through experience and observation. Master builders were those who had acquired that knowledge and the skills necessary for working with available materials, often through disastrous trial and error.

The role of materials changed dramatically with the advent of the Industrial Revolution. Rather than depending on an intuitive and empirical understanding of material properties and performance, architects began to be confronted with engineered materials. Indeed, the history of modern architecture can almost be viewed through the lens of the history of architectural materials. Beginning in the 19th century with the widespread introduction of steel, leading to the emergence of long-span and high-rise building forms, materials transitioned from their pre-modern role of being subordinate to architectural needs into a means to expand functional performance and open up new formal responses. The industrialization of glass-making coupled with developments in environmental systems enabled the 'international style' in which a transparent architecture could be sited in any climate and in any context. The broad proliferation of curtain wall systems allowed the disconnection of the façade material from the building's structure and infrastructure, freeing the material choice from utilitarian functions so that the façade could become a purely formal element. Through advancements in CAD/CAM (Computer Aided Design/Computer Aided Manufacturing) technologies, engineering materials such as aluminum and titanium can now be efficiently and easily employed as building skins, allowing an unprecedented range of building façades and forms. Materials have progressively emerged as providing the most immediately visible and thus most appropriable manifestation of a building's representation, both interior and exterior. As a result, today's architects often think of materials as part of a design palette from which materials can be chosen and applied as compositional and visual surfaces.

It is in this spirit that many have approached the use of smart materials. Smart materials are often considered to be a logical extension of the trajectory in materials development toward more selective and specialized performance. For many centuries one had to accept and work with the properties of a standard material such as wood or stone, designing to accommodate the material's limitations, whereas during the 20th century one could begin to select or engineer the properties of a high performance material to meet a specifically defined need. Smart materials allow even a further specificity – their properties are changeable and thus responsive to *transient* needs. For example, photochromic materials change their color (the property of spectral transmissivity) when exposed to light: the more intense the incident light, the darker the surface. This ability to respond to multiple states rather than being optimized for a single state has rendered smart materials a seductive addition to the design palette since buildings are always confronted with changing conditions. As a result, we are beginning to see many proposals speculating on how smart materials could begin to replace more conventional building materials.

Cost and availability have, on the whole, restricted widespread replacement of conventional building materials with smart materials, but the stages of implementation are tending to follow the model by which 'new' materials have traditionally been introduced into architecture: initially through highly visible showpieces (such as thermochromic chair backs and electrochromic toilet stall doors) and later through high profile 'demonstration' projects such as Diller and Scofidio's Brasserie Restaurant on the ground floor of Mies van der Rohe's seminal Seagram's Building. Many architects further imagine building surfaces, walls and façades composed entirely of smart materials, perhaps automatically enhancing their design from a pedestrian box to an interactive arcade. Indeed, terms like interactivity and transformability have already become standard parts of the architect's vocabulary even insofar as the necessary materials and technologies are far beyond the economic and practical reality of most building projects.

Rather than waiting for the cost to come down and for the material production to shift from lots weighing pounds to those weighing tons, we should step back and ask if we are ignoring some of the most important characteristics of these materials. Architects have conceptually been trying to fit smart materials into their normative practice alongside conventional building materials. Smart materials, however, represent a radical departure from the more normative building materials. Whereas standard building materials are static in that they are intended to *withstand* building forces, smart materials are dynamic in that they *behave* in response to energy fields. This is an important distinction as our normal means of representation in architectural design privileges the static material: the plan, section and elevation drawings of orthographic projection fix in location and in view the physical components of a building. One often designs with the intention of establishing an image or multiple sequential images. With a smart material, however, we should be focusing on what we want it do, not on how we want it to look. The understanding of smart materials must then reach back further than simply the understanding of material

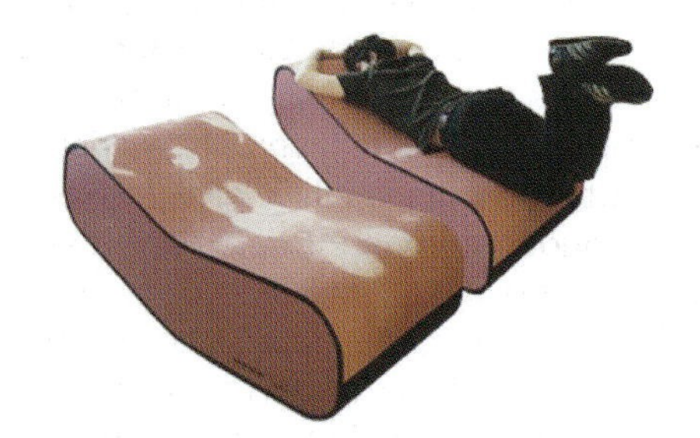

▲ **Figure 1-3** The 'heat' chair that uses thermochromic paint to provide a marker of where and when the body rested on the surface. (Courtesy of Juergen Mayer H)

properties; one must also be cognizant of the fundamental physics and chemistry of the material's interactions with its surrounding environment. The purpose of this book is thus two-fold: the development of a basic familiarity with the characteristics that *distinguish* smart materials from the more commonly used architectural materials, and speculation into the potential of these characteristics when deployed in architectural design.

1.2 The contemporary design context

Orthographic projection in architectural representation inherently privileges the surface. When the three-dimensional world is sliced to fit into a two-dimensional representation, the physical objects of a building appear as flat planes. Regardless of the third dimension of these planes, we recognize that the eventual occupant will rarely see anything other than the surface planes behind which the structure and systems are hidden. While the common mantra is that architects design space the reality is that architects make (draw) surfaces. This privileging of the surface drives the use of materials in two profound ways. First is that the material is identified as the surface: the visual understanding of architecture is determined by the visual qualities of the material. Second is that because architecture is synonymous with surface – and materials are that surface – we essentially think of materials as planar. The result is that we tend to consider materials in large two-dimensional swaths: exterior cladding, interior sheathing. Many of the materials that we do not see, such as insulation or vapor barriers, are still imagined and configured as sheet products. Even materials that form the three-dimensional infrastructure of the building, such as structural steel or concrete, can easily be represented through a two-dimensional picture plane as we tend to imagine them as continuous or monolithic entities. Most current attempts to implement smart materials in architectural design maintain the vocabulary of the two-dimensional surface or continuous entity and simply propose smart materials as replacements or substitutes for more conventional materials. For example, there have been many proposals to replace standard curtain wall glazing with an electrochromic glass that would completely wrap the building façade. The reconsideration of smart material implementation through another paradigm of material deployment has yet to fall under scrutiny.

One major constraint that limits our current thinking about materials is the accepted belief that the spatial envelope behaves like a boundary. We conceive of a room as a container of ambient air and light that is bounded or differentiated by its surfaces; we consider the building envelope to demarcate and separate the exterior environment from the interior environment. The presumption that the physical boundaries are one and the same as the spatial boundaries has led to a focus on highly integrated, multi-functional systems for façades as well as for many interior partitions such as ceilings and floors. In 1981, Mike Davies popularized the term 'polyvalent wall', which described a façade that could protect from the sun, wind and rain, as well as provide insulation, ventilation and daylight.[2] His image of a wall section sandwiching photovoltaic grids, sensor layers, radiating sheets, micropore membranes and weather skins has

▲ **Figure 1-4** Aerogel has a density only three times that of air, but it can support significant weights and is a superb insulator. Aerogels were discovered in 1931 but were not explored until the 1970s. (NASA)

influenced many architects and engineers into pursuing the 'super façade' as evidenced by the burgeoning use of double-skin systems. This pursuit has also led to a quest for a 'super-material' that can integrate together the many diverse functions required by the newly complex façade. Aerogel has emerged as one of these new dream materials for architects: it insulates well yet still transmits light, it is extremely lightweight yet can maintain its shape. Many national energy agencies are counting on aerogel to be a linchpin for their future building energy conservation strategies, notwithstanding its prohibitive cost, micro-structural brittleness and the problematic of its high insulating value, which is only advantageous for part of the year and can be quite detrimental at other times.

1.3 The phenomenological boundary

Missing from many of these efforts is the understanding of how boundaries physically behave. The definition of boundary that people typically accept is one similar to that offered by the Oxford English Dictionary: *a real or notional line marking the limits of an area*. As such, the boundary is static and defined, and its requirement for legibility (marking) prescribes that it is a tangible barrier – thus a visual artifact. For physicists, however, the boundary is not a thing, but an action. Environments are understood as energy fields, and the boundary operates as the transitional zone between different states of an energy field. As such, it is a place of change as an environment's energy field transitions from a high-energy to low-energy state or from one form of energy to another. Boundaries are therefore, by definition, active zones of mediation rather than of delineation. We can't see them, nor can we draw them as known objects fixed to a location.

Breaking the paradigm of the hegemonic 'material as visual artifact' requires that we invert our thinking; rather than simply visualizing the end result, we need to imagine the transformative actions and interactions. What was once a blue wall could be simulated by a web of tiny color-changing points that respond to the position of the viewer as well as to the location of the sun. Large HVAC (heating, ventilating and air conditioning) systems could be replaced with discretely located micro-machines that respond directly to the heat exchange of a human body. In addition, by investigating the transient behavior of the material, we challenge the privileging of the static planar surface. The 'boundary' is no longer delimited by the material surface, instead it may be reconfigured as the *zone in which change occurs.* The image of the building boundary as the demarcation between two different environments defined as single states – a homogeneous interior and an ambient exterior – could possibly be replaced by the idea of multiple energy environments fluidly interacting with the moving body. Smart materials, with their transient behavior and ability to respond to energy stimuli, may eventually enable the selective creation and design of an individual's sensory experiences.

Are architects in a position or state of development to implement and exploit this alternative paradigm, or, at the very least, to rigorously explore it? At this point, the answer is most probably no, but there are seeds of opportunity from on-going physical research and glimpses of the future use of the technology from other design fields. Advances in physics have led to a new understanding of physical phenomena, advances in biology and neurology have led to new discoveries regarding the human sensory system. Furthermore, smart materials have been comprehensively experimented with and rapidly adopted in many other fields – finding their way into products and uses as diverse as toys and automotive components. Our charge is to examine the knowledge gained in other disciplines, but develop a framework for its application that is suited to the unique needs and possibilities of architecture.

1.4 Characteristics of smart materials and systems

DEFINITIONS

We have been liberally using the term 'smart materials' without precisely defining what we mean. Creating a precise definition, however, is surprisingly difficult. The term is already in wide use, but there is no general agreement about what it actually means. A quick review of the literature indicates that terms like 'smart' and 'intelligent' are used almost interchangeably by many in relation to materials and systems, while others draw sharp distinctions about which qualities or capabilities are implied. NASA defines smart materials as 'materials that "remember" configurations and can conform to them when given a specific stimulus',[3] a definition that clearly gives an indication as to how NASA intends to investigate and apply them. A more sweeping definition comes from the *Encyclopedia of Chemical*

Technology: 'smart materials and structures are those objects that sense environmental events, process that sensory information, and then act on the environment'.[4] Even though these two definitions seem to be referring to the same type of behavior, they are poles apart. The first definition refers to materials as substances, and as such, we would think of elements, alloys or even compounds, but all would be identifiable and quantifiable by their molecular structure. The second definition refers to materials as a series of actions. Are they then composite as well as singular, or assemblies of many materials, or, even further removed from an identifiable molecular structure, an assembly of many systems?

If we step back and look at the words 'smart' and 'intelligent' by themselves we may find some cues to help us begin to conceptualize a working definition of 'smart materials' that would be relevant for designers. 'Smart' implies notions of an informed or knowledgeable response, with associated qualities of alertness and quickness. In common usage, there is also frequently an association with shrewdness, connoting an intuitive or intrinsic response. Intelligent is the ability to acquire knowledge, demonstrate good judgment and possess quickness in understanding.

Interestingly, these descriptions are fairly suggestive of the qualities of many of the smart materials that are of interest to us. Common uses of the term 'smart materials' do indeed suggest materials that have intrinsic or embedded quick response capabilities, and, while one would not commonly think about a material as shrewd, the implied notions of cleverness and discernment in response are not without interest. The idea of discernment, for example, leads one to thinking about the inherent power of using smart materials selectively and strategically. Indeed, this idea of a strategic use is quite new to architecture, as materials in our field are rarely thought of as performing in a direct or local role. Furthermore, selective use hints at a discrete response – a singular action but not necessarily a singular material. Underlying, then, the concept of the intelligent and designed response is a seamless quickness – immediate action for a specific and transient stimulus.

Does 'smartness', then, require special materials and advanced technologies? Most probably no, as there is nothing a smart material can do that a conventional system can't. A photochromic window that changes its transparency in relation to the amount of incident solar radiation could be replaced by a globe thermometer in a feedback control loop sending signals to a motor that through mechanical linkages repositions louvers on the surface of the glazing, thus changing the net transparency. Unwieldy, yes, but nevertheless feasible and possible to achieve with commonly used technology and materials. (Indeed, many buildings currently use such a system.) So perhaps the most unique aspects of these materials and technologies are the underlying concepts that can be gleaned from their behavior.

Whether a molecule, a material, a composite, an assembly, or a system, 'smart materials and technologies' will exhibit the following characteristics:

- Immediacy – they respond in real-time.
- Transiency – they respond to more than one environmental state.
- Self-actuation – intelligence is internal to rather than external to the 'material'.
- Selectivity – their response is discrete and predictable.
- Directness – the response is local to the 'activating' event.

It may be this last characteristic, directness, that poses the greatest challenge to architects. Our building systems are neither discrete nor direct. Something as apparently simple as changing the temperature in a room by a few degrees will set off a Rube Goldberg cascade of processes in the HVAC system, affecting the operation of equipment throughout the building. The concept of directness, however, goes beyond making the HVAC equipment more streamlined and local; we must also ask fundamental questions about the intended behavior of the system. The current focus on high-performance buildings is directed toward improving the operation and control of these systems. But why do we need these particular systems to begin with?

The majority of our building systems, whether HVAC, lighting, or structural, are designed to service the building and hence are often referred to as 'building services'. Excepting laboratories and industrial uses, though, buildings exist to serve their occupants. Only the human body requires management of its thermal environment, the building does not, yet we heat and cool the entire volume. The human eye perceives a tiny fraction of the light provided in a building, but lighting standards require constant light levels throughout the building. If we could begin to think of these environments at the small scale – what the body needs – and not at the large scale – the building space – we could dramatically reduce the energy and material investment of the large systems while providing better conditions for the human occupants. When these systems were conceived over a century ago, there was neither the technology nor the knowledge to address human needs in any manner other than through large indirect systems that provided homogeneous building conditions. The advent of smart materials now enables the design of direct and discrete environments for the body, but we have no road map for their application in this important arena.

1.5 Moving forward

Long considered as one of the roadblocks in the development and application of smart materials is the confusion over which discipline should 'own' and direct the research efforts as well as oversee applications and performance. Notwithstanding that the 'discovery' of smart materials is attributed to two chemists (Jacques and Pierre Curie no less!), the disciplines of mechanical engineering and electrical engineering currently split ownership. Mechanical engineers deal with energy stimuli, kinematic (active) behavior and material structure, whereas electrical engineers are responsible for microelectronics (a fundamental component of many smart systems and assemblies), and the operational platform (packaging and circuitry). Furthermore, electrical engineers have led the efforts toward miniaturization, and as such, much of the fabrication, which for most conventional materials would be housed in mechanical engineering, is instead under the umbrella of electrical engineering.

This alliance has been quite effective in the development of new technologies and materials, but has been less so in regard to determining the appropriate applications. As a result, the smart materials arena is often described as 'technology push' or, in other words, technologies looking for a problem. Although this is an issue that is often raised in overviews and discussions of smart materials, it has been somewhat nullified by the rapid evolution and turnover of technologies in general. Many industries routinely adopt and discard technologies as new products are being developed and old ones are upgraded. As soon as knowledge of a new smart material or technology enters the public realm, industries of all sizes and of all types will begin trying it out, eliminating the round pegs for the square holes. This trial and error of matching the technology to a problem may well open up unprecedented opportunities for application that would have gone undetected if the more normative 'problem first' developmental sequence had occurred. For architecture, however, this reversal is much more problematic.

In most fields, technologies undergo continuous cycles of evolution and obsolescence as the governing science matures;

as a result, new materials and technologies can be easily assimilated. In architecture, however, technologies have very long lifetimes, and many factors other than science determine their use and longevity. There is no mechanism by which new advances can be explored and tested, and the small profit margin in relation to the large capital investment of construction does not allow for in situ experimentation. Furthermore, buildings last for years – 30 on average – and many last for a century or more. In spite of new construction, the yearly turnover in the building stock is quite low. Anything new must be fully verified in some other industry before architects can pragmatically use it, and there must also be a match with a client who is willing to take the risk of investing in any technology that does not have a proven track record.

The adoption of smart materials poses yet another dilemma for the field of architecture. Whereas architects choose the materials for a building, engineers routinely select the technologies and design the systems. Smart materials are essentially material systems with embedded technological functions, many of which are quite sophisticated. Who, then, should make the decisions regarding their use? Compounding this dilemma are the technologies at the heart of smart materials; the branches of mechanical and electrical engineering responsible for overseeing this area have virtually no connection to or relationship with the engineering of building systems. Not only are smart materials a radical departure from the more normative materials in appearance, but their embedded technology has no precedent in the large integrated technological systems that are the standard in buildings.

How can architects and designers begin to explore and exploit these developing technologies and materials, with the recognition that their operating principles are among the most sophisticated of any technologies in use? Although architecture is inherently an interdisciplinary profession, its practice puts the architect at the center, as the director of the process and the key decision-maker. The disciplines that we must now reach out to, not only mechanical and electrical engineering, but also the biological sciences, have little common ground. There are no overlapping boundaries in knowledge, such as you might find between architecture and urban design, and there is no commonality of problem, such as you might find between architecture and ecology. Our knowledge base, our practice arena, and even our language are split from those in the smart materials domain. Ultimately, our use of these materials will put us into the heady role of manipulating the principles of physics.

1.6 Organization of the text

The objectives of this book are thus three-fold. The first is to provide a primer on smart materials, acquainting architects and designers with the fundamental features, properties, behaviors and uses of smart materials. Of particular importance is the development of a vocabulary and a descriptive language that will enable the architect to enter into the world of the material scientist and engineer. The second objective is the framing of these new materials and technologies as behaviors or actions and not simply as artifacts. We will be describing smart materials in relation to the stimulus fields that surround them. Rather than categorizing materials by application or appearance, we will then categorize them in relation to their actions and their energy stimulus. Our third objective is the development of a methodological approach for working with these materials and technologies. We will successively build systems and scenarios as the book progresses, demonstrating how properties, behaviors, materials and technologies can be combined to create new responses. If these three objectives are met, the designer will be able to take a more proactive stance in determining the types of materials and systems that should be developed and applied. Furthermore, competency in the foundations of energy and material composition behavior will eventually allow the architect or designer to think at a conceptual level 'above' that of the material or technology. One of the constants in the field of smart materials is that they are continuously being updated or replaced. If we understand classes of behaviors in relation to properties and energy fields, then we will be able to apply that understanding to any new material we may 'meet' in the future.

To pull these objectives together, the overall organization of the book follows a bipartite system; categories of behavior will be established and then will be overlaid with increasing component and system complexity. Chapter 2 serves as the entry into the subject of material properties and material behavior, whereas Chapter 3 first posits the framework through which we will categorize smart materials. We will establish a basic relationship between material properties, material states and energy that we can use to describe the interaction of all materials with the environments – thermal, luminous and acoustic – that surround the human body. This basic relationship forms a construct that allows us to understand the fundamental mechanisms of 'smartness'. The resulting construct will form the basis not only for the categories, but will also be useful as we discuss potential combinations and applications.

Smartness in a material or system is determined by one of two mechanisms, which can be applied directly to a singular material, and conceptually to a compound system (although individual components may well have one of the direct mechanisms). If the mechanism affects the internal energy of the material by altering either the material's molecular structure or microstructure then the input results in a property change of the material. (The term 'property' is important in the context of this discussion and will be elaborated upon later. Briefly, the properties of a material may be either intrinsic or extrinsic. Intrinsic properties are dependent on the internal structure and composition of the material. Many chemical, mechanical, electrical, magnetic and thermal properties of a material are normally intrinsic to it. Extrinsic properties are dependent on other factors. The color of a material, for example, is dependent on the nature of the external incident light as well as the micro-structure of the material exposed to the light.) If the mechanism changes the energy state of the material, but does not alter the material *per se*, then the input results in an exchange of energy from one form to another. A simple way of differentiating between the two mechanisms is that for the property change type (hereafter defined as Type I), the material absorbs the input energy and undergoes a change, whereas for the energy exchange type (Type II), the material stays the same but the energy undergoes a change. We consider both of these mechanisms to operate directly at the micro-scale, as none will affect anything larger than the molecule, and furthermore, many of the energy-exchanges take place at the atomic level. As such, we cannot 'see' this physical behavior at the scale at which it occurs.

▲ **Figure 1-5** Radiant color film. The color of the transmitted or reflected light depends upon the vantage point. Observers at different places would see different colors (*see* Chapter 6)

HIGH-PERFORMANCE VERSUS SMART MATERIALS

We will soon begin to use the construct just described to begin characterizing smart materials, and specifically look at materials that change their properties in response to varying external stimuli and those that provide energy transformation functions. This construct is specific to smart materials. It does not reflect, for example, many extremely exciting and useful new materials currently in vogue today. Many of these interesting materials, such as composites based on carbon fibers or some of the new radiant mirror films, change neither their properties nor provide energy transfer functions; and hence are not smart materials. Rather, they are what might best be described as 'high-performance' materials. They often

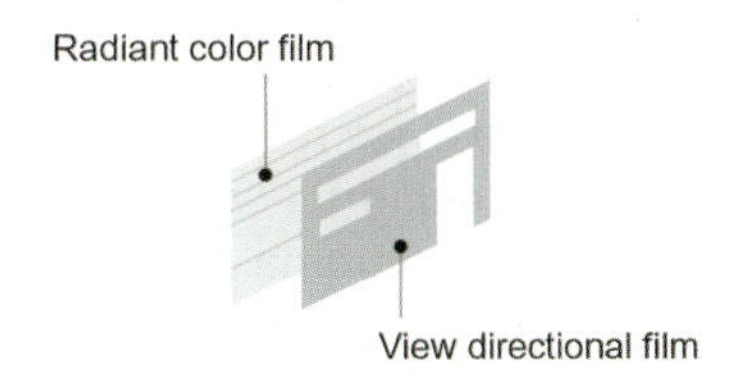

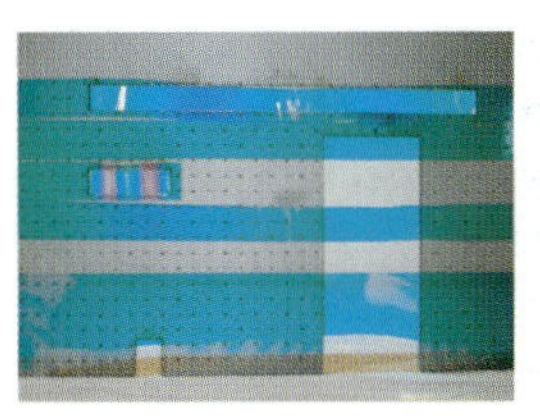

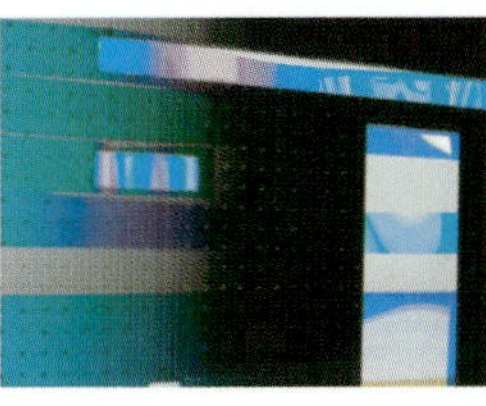

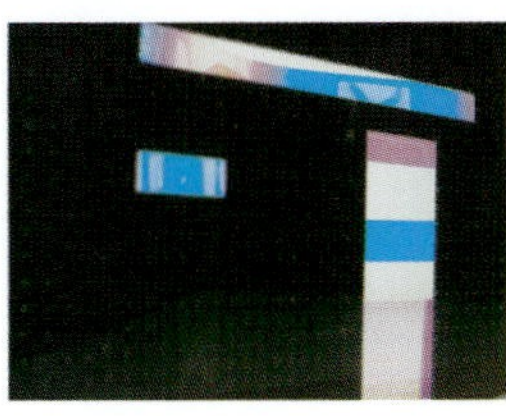

▲ **Figure 1-6** Design experiment: view directional film and radiant color film have been used together in this façade study. (Nyriabu Nyriabu)

have what might be called 'selected and designed properties' (e.g., extremely high strength or stiffness, or particular reflective properties). These particular properties have been optimized via the use of particular internal material structures or compositions. These optimized properties, however, are static. Nevertheless, we will still briefly cover selected high performance materials later in Chapter 4 because of the way they interact with more clearly defined smart materials.

TYPE 1 MATERIALS

Based on the general approach described above, smart materials may be easily classified in two basic ways. In one construct we will be referring to materials that undergo changes in one or more of their properties – chemical, mechanical, electrical, magnetic or thermal – in direct response to a change in the external stimuli associated with the environment surrounding the material. Changes are direct and reversible – there is no need for an external control system to cause these changes to occur. A photochromic material, for example, changes its color in response to a change in the amount of ultraviolet radiation on its surface. We will be using the term 'Type 1' materials to distinguish this class of smart materials.

Chapter 4 will discuss these materials in detail. Briefly, some of the more common kinds of Type 1 materials include the following:

- Thermochromic – an input of thermal energy (heat) to the material alters its molecular structure. The new molecular structure has a different spectral reflectivity than does the original structure; as a result, the material's 'color' – its reflected radiation in the visible range of the electromagnetic spectrum – changes.
- Magnetorheological – the application of a magnetic field (or for electrorheological – an electrical field) causes a

change in micro-structural orientation, resulting in a change in viscosity of the fluid.

- Thermotropic – an input of thermal energy (or radiation for a phototropic, electricity for electrotropic and so on) to the material alters its micro-structure through a phase change. In a different phase, most materials demonstrate different properties, including conductivity, transmissivity, volumetric expansion, and solubility.
- Shape memory – an input of thermal energy (which can also be produced through resistance to an electrical current) alters the microstructure through a *crystalline* phase change. This change enables multiple shapes in relationship to the environmental stimulus.

▲ **Figure 1-7** A 'cloth' made by weaving fiber-optic strands that are lighted by light-emitting diodes (LEDs). (Yokiko Koide)

TYPE 2 MATERIALS

A second general class of smart materials is comprised of those that transform energy from one form to an output energy in another form, and again do so directly and reversibly. Thus, an electro-restrictive material transforms electrical energy into elastic (mechanical) energy which in turn results in a physical shape change. Changes are again direct and reversible. We will be calling these 'Type 2' materials. Among the materials in this category are piezoelectrics, thermoelectrics, photovoltaics, pyroelectrics, photoluminescents and others.

Chapter 4 will also consider these types of materials at length. The following list briefly summarizes some of the more common energy-exchanging smart materials.

- Photovoltaic – an input of radiation energy from the visible spectrum (or the infrared spectrum for a thermo-photovoltaic) produces an electrical current (the term voltaic refers more to the material which must be able to provide the voltage potential to sustain the current).
- Thermoelectric – an input of electrical current creates a temperature differential on opposite sides of the material. This temperature differential produces a heat engine, essentially a heat pump, allowing thermal energy to be transferred from one junction to the other.
- Piezoelectric – an input of elastic energy (strain) produces an electrical current. Most piezoelectrics are bi-directional in that the inputs can be switched and an applied electrical current will produce a deformation (strain).
- Photoluminescent – an input of radiation energy from the ultraviolet spectrum (or electrical energy for an electroluminescent, chemical reaction for a chemoluminescent) is converted to an output of radiation energy in the visible spectrum.
- Electrostrictive – the application of a current (or a magnetic field for a magnetostrictive) alters the inter-atomic distance through polarization. A change in this distance changes the energy of the molecule, which in this case produces elastic energy – strain. This strain deforms or changes the shape of the material.

With Type 2 materials, however, we should be aware that use of the term 'material' here can be slightly misleading. Many of the 'materials' in this class are actually made up of several more basic materials that are constituted in a way to provide a particular type of function. A thermoelectric, for example, actually consists of multiple layers of different materials. The resulting assembly is perhaps better described as a simple device. The term 'material', however, has still come to be associated with these devices – largely because of the way they are conceptually thought about and used. Application-oriented thinking thus drives use of the term 'material' here.

SMART SENSORS, ACTUATORS AND CONTROL SYSTEMS

Compounding the problematic of terminology, we will see that many smart materials may also inherently act as sensors or actuators. In their role as sensors, a smart material responds to a change in its environment by generating a perceivable response. Thus, a thermochromic material could be used directly as a device for sensing a change in the temperature of an environment via its color response capabilities. Other materials, such as piezoelectric crystals, could also be used as actuators by passing an electric current through the material to create a force. Many common sensors and actuators are based on the use of smart materials.

In the use of Type 2 materials as a sensor or actuator, there are also different kinds of electronic systems that are integral to the system to amplify, modify, transmit, or interpret generated signals. Logic capabilities provided via microprocessors or other computer-based systems are similarly common. There are several different types of strategies possible here. We will return to this topic in Chapter 5.

COMPONENTS AND SYSTEMS

As is common in any design context, basic types of smart materials are normally used in conjunction with many other materials to produce devices, components, assemblies and/or systems that serve more complex functions. As was previously mentioned, external walls in a building, for example, provide a range of pragmatic functions (thermal barrier, weather enclosure, ventilation, etc.) as well as establishing the visual experience of a building. Single materials cannot respond to these many demands alone. Thus, we might have a whole series of different types of 'smart walls' depending on exactly how the wall is constituted, what primary functions it is intended to serve and the degree to which there are external logic controls.

In addition to constructions that we normally think of as components, we also have whole systems in buildings that can be designed to possess some level of smartness. The

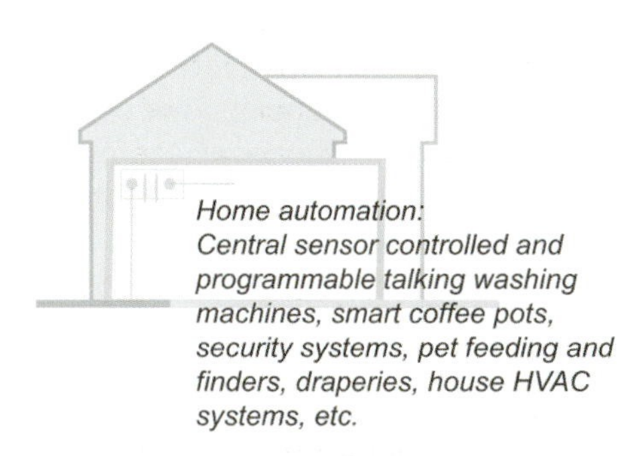

'Smart Home' - Current versions of smart homes depend on automation and information technology

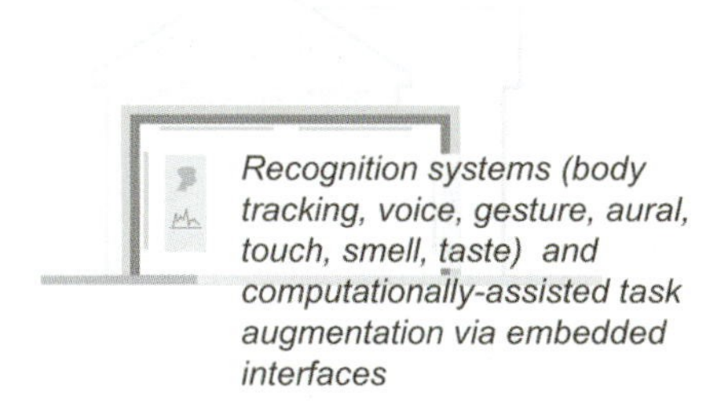

Current 'intelligent rooms' depend on information-rich multimodal environments that are context-aware of occupants

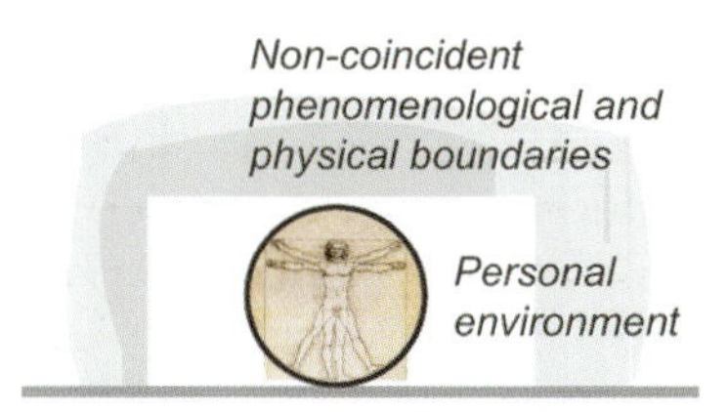

Future approaches will feature increasing cognition and context-aware response levels suggestive of biological models, but may also see an evolution of the personal environment and a devolution of traditional physical and phenomenological boundaries

▲ **Figure 1-8** Current smart room and intelligent room paradigms, with a glimpse into the future (*see* Chapter 8)

systems of concern here include normal environmental systems (heating, ventilation and air conditioning; lighting; acoustical) and structural systems. Historically, one of the first uses of the term 'smart' was in connection with improved sensor-based monitoring and control systems for controlling the thermal environment in a building (the 'Smart House' of the 1990s). Whether or not this approach is commensurate with the term 'smart' as it is used today is an interesting question, and one that we will return to in Chapter 7. In that chapter we will consider different kinds of smart systems in use today.

SMART VS INTELLIGENT ENVIRONMENTS

Fundamentally, the product of architecture and design is a complete work – whether a building or a lamp. Inherent to each, however, is a stunning complexity in all of its aspects. Here the question is naturally raised of the notion of smart and/or intelligent environments. The term 'intelligent' itself is as problematic as the term 'smart', yet it surely suggests something of a higher level than does 'smart'. We do expect more out of 'intelligent systems' than we do from 'smart materials'. Everyday connotations of the term 'intelligent' with suggested notions of abilities to understand or comprehend, or having the power of reflection or reason, could be useful, and will help us as we examine the current conceptions of these environments and develop new ones of our own.

One of the more fascinating aspects of today's society is how 'techno-speak' terms come into existence and assume currency without universal agreement about what is actually meant. There has been a lot of recent interest in 'intelligent rooms' and 'intelligent buildings' without a clear consensus about what is actually meant by these terms. The parallel question raised of whether common rooms or buildings are 'dumb' is equally interesting, particularly since architects and builders have done rather well at responding to societal and cultural needs for millennia. More specific fundamental needs have not been ignored, nor have the wonderful vicissitudes of human desire. So, presumably, something else and more specific is meant by the terms 'intelligent rooms' or 'intelligent buildings', but what? Here we also engage in another meaning conundrum. The phrase 'smart environments' is in widespread use and has already been employed in this book. What, if anything, is the difference between an environment or building space that is 'intelligent' and one that is 'smart?' The engineering and computer science worlds often do not distinguish between the two, presuming that both represent the crowning culmination of technological development – that of the fully contained and controlled environment. In Chapters 8 and 9, we begin to propose an alternative in which systems become smaller and more discrete, freeing our bodies and our environments from an overarching web of control. It is perhaps in this arena that architects can have the most impact on the trajectory of these advanced materials and technologies.

Notes and references

1 All discussion on alchemy in this chapter is from David. C. Lindberg (ed.), *Science in the Middle Ages* (Chicago: The University of Chicago Press, 1978). See in particular chapter 11 on the 'Science of Matter'.

2 Davies, M. (1981) 'A wall for all seasons', *RIBA Journal*, 88 (2), pp. 55–57. The term 'polyvalent wall', first introduced in this article, has become synonymous with the 'advanced façade' and most proposals for smart materials in buildings are based on the manifestation of this 1981 ideal.

3 http://virtualskies.arc.nasa.gov/research/youDecide/smartMaterials.html.

4 Kroschwitz, J. (ed.) (1992) *Encyclopedia of Chemical Technology*. New York: John Wiley & Sons.

2 Fundamental characterizations of materials

Chapter 1 provided a brief insight into how smart materials and systems might affect our design thinking. We identified five 'conceptual' characteristics – immediacy, transiency, self-actuation, selectivity and directness – that differentiated these materials from more traditional materials, but we need further information regarding the 'physical' characteristics of these materials and technologies. For example, how do we measure transiency? Or what defines discretion in an assembly? As designers, we understand conceptual characteristics as intentions, whereas engineers understand physical characteristics as tools in implementation. This chapter will begin to lay out the tangible definitions that are necessary for bridging this gap in knowledge and in application. In order to achieve this we must directly address the question of how best to classify these materials. Classification systems and related taxonomies are useful not only for simple categorization and description purposes, but they can invariably suggest more far-reaching fundamental constructs of a field. This precept is particularly important in our current context in which the smart materials field is just emerging. Examining the structure of different classifications for materials will help us to place smart materials within a broader context. After reviewing several existing and common approaches for classification, we will develop our own structure that marries the intentions of the architect with the tools of the engineer.

The latter half of this chapter will provide the necessary overview of materials science, beginning with atomic structure and concluding with material properties. Fundamental to the development of a new construct for the exploitation of these materials in the design professions is an understanding of the origin and determinants of their behaviors. Just as the responses of these materials are discrete and direct, then our interaction with them must ultimately function at the same scale, whether atomic, molecular, or micro-structural. Designers are used to manipulating materials at the object scale, and while a large-scale interaction will invariably impact smaller-scale behavior, we can operate more efficiently, predictably and quickly if we act directly on the root mechanism of the behavior.

2.1 Traditional material classification systems

There are a number of existing classification and descriptive systems used in connection with materials. One broad approach stems from a fairly basic materials science approach to the subject matter, wherein the primary point of view revolves around the internal structure of the material. Another approach commonly used in the engineering profession is essentially descriptive but focuses on the performance characteristics of materials. In the design fields, a host of different loose categorizations are used, many of which are particular (and perhaps idiosyncratic) to individual fields. For example, interior designers maintain classification types that are distinctly different from those used in landscape architecture. There are also various kinds of classifications that literally provide the legal basis for the specification of materials in design works. In general, we will see that each material system adopts a particular point of view that is useful to a particular construct of the field and/or for a particular application. The construct may have no overlap or applicability for another group. Hence, it is important to understand these points of view.

MATERIAL SCIENCE CLASSIFICATIONS

The material science approach to classification goes directly to the core understanding of the basic internal structure of materials. As a result, we might consider this system to be *compositionally* driven. The most fundamental level of differentiation begins with the bonding forces between individual atoms. It is this bonding force, whether ionic, covalent, metallic or Van der Waals, that will ultimately determine many of the intrinsic properties and major behavioral differences between materials. The next level of description hinges on the way these bonding forces produce different types of aggregation patterns between atoms to form various molecular and crystalline solid structures. These larger aggregation patterns can further be differentiated by how their molecular structures branch or link or, in crystalline solids, by different types of unit cell and related spatial lattice structures such as face-centered or body-centered. Diamond, for example, is a covalently bonded material with a cubic crystal structure. At the highest level are the broadly descriptive categories such as ceramics, metals or polymers, which are familiar to us even insofar as the boundaries between these classes are not nearly as distinct as at the lower levels of the classification system – silicones exist between ceramics and plastics, and many semiconductors could be either a metal or ceramic.

This way of classifying materials is extremely useful for many reasons. In particular, the understandings reflected in the classifications provide a way of describing the specific qualities or properties (e.g., hardness, electrical conductivity) that characterize different materials. Knowledge of properties at the atomic and molecular level can transform our impression of smart materials from 'gee-whiz' materials into an understanding of them through scientifically described attributes and behaviors. Consequently, it also provides a basis for developing a method for designing materials that possess different qualities or properties.

ENGINEERING CLASSIFICATIONS

Applied classification approaches are shown in Figures 2–1 and 2–2. These types are primarily used in the mechanical engineering profession to distinguish between the fundamental *problem-solving* characteristics of the nearly 300 000 materials readily available to the engineer (steel alone has over 2000 varieties). Rather than the hierarchical organization of the material scientist, the engineering classification is one of mapping, enabling the engineer to mix and match properties and attributes to best solve the problem or need at hand. Materials in the engineering realm are chosen based on what they can do, how they behave and what they can withstand. The physical criteria for the use are first determined, and a material is selected or engineered to provide the best fit or, at the very least, the most acceptable compromise for the specific criteria. If the material science classification describes how a material is composed, then the engineering classification explains what it does. Furthermore, since the focus is always a practical one, i.e. the material will be used in a product or process, many of the categories are quite pragmatic. For example, an important category is the environment of the application: can the material function in a corrosive atmosphere, can it withstand being submerged in sea water? Still other engineering classifications might include cost, availability, or recyclability as categories of equal importance to the more basic descriptive ones such as state and composition. Even though the final objective in all engineering applications is the optimization of a material property for a particular situation, regardless of the material type, the additional criteria will quickly narrow down the

STATE	solid, liquid, gas
STRUCTURE	amorphous, crystalline
ORIGIN	natural, synthetic
COMPOSITION	organic, inorganic, alloy
PROCESSING	cast, hardened, rolled
PROPERTY	emissivity, conductivity
ENVIRONMENT	corrosive, underwater
APPLICATION	adhesive, paint, fuel

▲ **Figure 2-1** Basic organization of material catgeories in the engineering profession with a few examples in each category. Engineers must weigh many of these characteristics in choosing a material. (Adapted from Myer P. Kutz (ed.), *The Mechanical Engineer's Handbook*. New York: John Wiley, 1998)

Engineering materials

Group	Class	Type	Examples
Ferrous metals	Steels	Carbon - low alloy steels High alloy steels	
	Cast irons	Gray, white, malleable cast iron Ductile iron	
Combination metals		Clad metals, coated metals, other	
Nonferrous metals	Engineering metals	Light metals	Aluminum/alloys Titanium/alloys Other
		Medium metals	Chromium/alloys Copper/alloys Other
		Heavy metals	Tin/alloys Zinc/alloys Other
	Specialty metals	Semiconductor, other	
Combination materials	Composites	Particle composites Fiber composites Dispersion composites	
	Foams, other	Foams, microspheres	
	Laminates	Clad laminates Bonded laminates Honeycomb laminates	
Nonmetals and compounds	Crystalline nonmetals	Minerals Ceramics	Refractory Nonrefractory
	Fibrous materials	Wood/products	Natural Treated Processed
		Textile fiber products	Natural Synthetic
	Amorphous materials	Glasses Plastics Rubber/elastomers	Thermoplastics Thermosets

▲ **Figure 2-2** This classification system for materials is typical of those used in applied engineering. It readily mixes the form of material structures (e.g., laminates, amorphous) with properties (ferrous, nonferrous), but can be very useful for many applications. It is difficult to use this kind of classification, however, to describe smart materials with property-changing or energy-exchanging characteristics. (Diagram modeled after Fig. 31-9 in Myer P. Kutz (ed.), *The Mechanical Engineer's Handbook*. New York: John Wiley, 1998) seemingly endless choices. Many industries have developed their own classification systems to help narrow down the choice of materials to those that are appropriate for their own uses. For example, the American Iron and Steel Institute, which deals only with ferrous materials, adopted a straightforward numbering system that encompasses alloy composition, carbon content and processing method. The American Welding Association is even more specific, categorizing electrode materials by tensile strength, welding technique and position. Regardless of the source of the classification system, each one clearly highlights properties that underpin the useful behavior of the material.

The behavior focus of the engineering classification is not as likely to lead toward the direct development of new materials as would be supported by the more compositionally focused system of the materials scientist. Nevertheless, by working toward the optimization of a property, rather than of a material, this focus on behavior is friendlier toward new materials. Desired behaviors, as defined by material properties, have no preference for specific materials or technologies, and, as a result, will be more suitable for and more open to experimentation and novel solutions.

TRADITIONAL ARCHITECTURAL CLASSIFICATIONS

There are several material classification approaches that have evolved over the years for describing the materials used in architectural settings. Many have a mix of classification perspectives, and are rarely based on pure performance requirements, as would be characteristic in engineering fields. Architectural building codes and standards, for example, often supersede performance criteria in an attempt to simplify the selection process and remove liability for performance failures. For many uses, codes and standards often explicitly or implicitly identify acceptable materials, leaving the architect only to select between brands. As a result, architectural classifications tend to be more nominative – simply listing materials and uses in accordance with standard building requirements.

Within architectural practice, these various requirements are codified in different ways. In the United States, the Construction Specifications Institute has maintained a standardized classification system for over 50 years. This system, known as the CSI index, organizes materials in two ways. The first places the materials typically used in a building into broad classes. In this section, we will find generic material groupings such as paint, laminate and concrete. The second organizes

The CSI Master Format	
Division 1	General Requirements
Division 2	Sitework
Division 3	Concrete
Division 4	Masonry
Division 5	Metals
Division 6	Wood and Plastics
Division 7	Thermal-Moisture Protection
Division 8	Doors and Windows
Division 9	Finishes
Division 10	Specialties
Division 11	Equipment
Division 12	Furnishings
Division 13	Special Construction
Division 14	Conveying Systems
Division 15	Mechanical
Division 16	Electrical

Division 8	Doors and Windows
08100	Metal doors and windows
08200	Wood and plastic doors
08250	Door opening assemblies
08300	Special doors
08400	Entrances and storefronts
08500	Metal windows
08600	Wood and plastic windows
08650	Special windows
08700	Hardware
08800	Glazing
08900	Glazed curtain walls

▲ **Figure 2-3** The Construction Specifications Institute (CSI) Master Format is a standard outline for construction specifications in the United States. To illustrate the depth of this format, Division 8 is presented in its expanded form

by component or system. These categories are equally generic and, furthermore, are not even material-specific. For example, windows fall into this category, even though they may be manufactured with wood, vinyl, aluminum or steel. The emphasis in both major groupings is toward application and common use; the fundamental behaviors and properties are incidental. In the broad material classes, the properties, performance and behavior are largely presumed to be satisfactory as long as the chosen material fits within the normative uses defined by practice. For example, the characteristics of wood are discussed in relationship to their relevance for the intended application: the grade of wood suitable for load-bearing roof structures, or the type of wood suitable for finish flooring. The system or component classes focus on application as well. Doors are organized according to their suitability for security, fire protection, egress, as well as by their use for commercial or residential buildings.

The CSI index also addresses the technologies typically used in buildings, grouping them into operational systems, such as heating, ventilating and air conditioning (HVAC), lighting and plumbing, and into constructional systems, such as structural, drainage and vertical circulation. This too differs substantially from the method for categorizing technologies in the engineering fields in which technologies are routinely organized by their process – e.g. smelting or CAD/CAM – or by their mechanism of operation – compression or pumping. Fewer specifics are made available to architects on these systems, as it is presumed that an engineer will be responsible for selecting building technologies.

Essentially, if the materials science classification explains '*why* one material is differentiated from another', and the engineering classification determines '*how* a material performs', then the architectural classifications operates at the other end of the sequence by listing '*what* a material is and *where* it is used'. This system is intended to remove the decision-making responsibility from the architect, and as such, it is less about informed choice and optimization and more about specification and standardization. The result is information, not knowledge.

This general approach is often the framework into which applications of new materials in architecture are forced to fit, and it has clearly proved problematic in this regard. Classification systems such as that of the CSI are not intended to spur innovation in the materials field. Rather, they are practical templates for communication between architects, contractors, fabricators and suppliers. After the preliminary design of a building is completed and approved, architects prepare construction documents that serve as the 'instructions' for the construction of the building. A textual document defines each building element on the design drawings and specifies the material or component. This document, rather than providing guidelines, instead serves as a binding contract that construction professionals and contractors must follow. Trade associations and manufacturers of building products routinely write their material and product specifications in this format to streamline the specification process for architects, and many architectural firms maintain an internal set of construction specifications that are used as the baseline for all of their projects. While communication and contractual applications are important, particularly in a field that has direct responsibility for the public's safety and welfare, the peripheral consequences of a specification-driven system generally result in the exclusion of new and unusual materials and technologies.

2.2 Alternative classification systems

Nevertheless, there have been many attempts to introduce new materials to designers through alternative classification systems. Many are quite qualitative and readily mix approaches to description, but almost all invert the criteria-driven process that characterize the materials science and engineering systems. In many design fields the material is chosen long before performance criteria are defined and as such the process tends to be artifact-driven. The rationale for this comes from many fronts, not all of which are based on physical requirements reflected in the mechanical engineering classification approach or the internal structuring of the material science perspective.

A good example of this general approach is reflected in the book *Technotextiles*, which converges on a specific subset of materials intended for use in the fashion design profession. Terms such as Fibres and Fabrics, Electronic Textiles, Engineered Textiles and Textile Finishes are used broadly to characterize the materials described in the book.[1] These categories are used to describe and illustrate many of the textiles used in the fashion design industry. Several descriptions are by the finishing process (e.g., Heat-transfer, Ink-Jet), some by general composition or form (Laminates, Composites), several by broad designations of material type (Glass, Metal), others by use (Coatings) and still others by geometry (Three-dimensional textiles). Product designers are similarly familiar with Mike Ashby's 'bubble charts', which visually represent material groups and their properties.[2]

To the engineer or scientist, there appears to be no common thread present in this descriptive system, yet it has been very useful to the fashion designer. The thread that is present is not a science-based understanding of the materials described; rather the approach touches on the information needed by the working fashion designer in selecting and using materials – a process in which materials are usually chosen on the basis of certain aesthetic qualities readily understood by the designer (with performance requirements considered afterwards). Current process orientations (e.g., ink-jet), for example, are known to produce particular kinds of visual characteristics known *a priori* to the designer. In this sense, the free mixing of perspectives can be useful and valuable. Nevertheless, a highly problematic aspect of this approach is that it is based almost entirely on current or past practices and thus further codifies them. This approach is also not useful to other groups important to the future of the field, such as to the materials scientist seeking to develop a new kind of polymer that exhibits specific mechanical properties, or to the mechanical engineer seeking to identify a material for use in a product such as an automobile body where performance requirements are paramount.

Material ConneXion®, a material library and resource bank in New York and Milan, attempts to circumvent the resistance to new material adoption in many of the design fields by including only unusual or novel materials in their collection. Most of the 3000+ materials in their collection are unprecedented in architecture, as they come from fields and applications with little crossover. For example, there are ceramic tiles used for furnace refractory lining, and polyamid resins for injection molding. The materials are organized similarly to the broad composition categories that sit at the top of the material science classification system, but are without the inductive lower layers that serve to explain the material. The eight broad categories – polymers, glass, ceramics, carbon-based, cement-based, metals, natural materials and natural material derivatives – also have little in common with the more normative architecture categories. While this is intended to break the hegemony of the currently over-specified process of material selection that abounds in the design fields, there is little contextualization of the categories. For example, the term polymer is not associated either with a familiar product or a particular use. Without an understanding of material behavior and structure, architects and designers fall back to a more familiar mode – choosing a material based on its visual characteristics.

If in the traditional engineering approach the material is understood as an array of physical behaviors, then in the traditional architectural and general design approaches the material is still conceived as a singular static thing, an artifact. Considering smart materials as fixed artifacts is clearly unsatisfactory as this neglects their contingency on their environment (their properties respond to and vary with external stimuli). The engineering approach is little better as it is based on a specificity of performance optimized to a single state that inherently denies the mutability of the material and its interactions with its surroundings. As a result, many of the materials and technologies that we are interested in have not been suitably categorized by other systems, including those of the engineering field.

2.3 Classification systems for advanced and smart materials

The information necessary for the implementation of new materials may be available, but there is as yet no method for its application in the design fields. Staying with the current method and treating smart materials as artifacts in a classification system is clearly problematic. Even if a smart material could be considered as a replacement for a conventional material in many components and applications, its inherent 'active' behavior makes it also potentially applicable as a technology. For example, electrochromic glass can be simultaneously a glazing material, a window, a curtain wall system, a lighting control system or an automated shading system. In the normative classification the product would then fall into several separate categories rendering it particularly difficult for the architect to take into consideration the multi-modal character and performance of the material. Furthermore, many of the new technologies are unprecedented in application, and thus have no place-holder in conventional descriptions.

Perhaps most fitting, then, is for smart material classifications to be multi-layered – with one layer characterizing the material according to its physical behavior (what it does) and another layer characterizing the material according to its phenomenological behavior (the results of the physical behavior). Phenomenological behavior is rarely documented, much less considered, in the field of architecture. We can categorize these effects in terms of their arena of action, which could be considered as analogous to an architect's intention – what do we want the material to do? The smart materials that we use can produce direct effects on the energy environments (luminous, thermal and acoustic), or they can produce indirect effects on systems (energy generation, mechanical equipment). This approach is operationally very useful to the designer in evaluating the use of smart materials and systems in relation to the design of environments.

We must also recognize, however, that there is both value and reality in considering how these materials are invariably used in the service of making ever-more complex devices, assemblies and environments that are intrinsically multi-modal or otherwise provide more complex responses than are possible with single materials. This is essentially a functions/systems approach. As noted in Chapter 1, this book follows a bi-partite approach: materials and technologies are categorized by behavior – both physical and phenomenological – and then overlaid with increasing component and system complexity. This layer enables us to meet and confront related new initiatives and technologies that shape larger devices and environments – especially those initiatives on 'intelligent environments' that spring primarily from the computational world. Here we must address questions previously raised about how smart materials relate to the world of intelligent devices and environments. As a way of structuring subsequent inquiries and discussions, a working

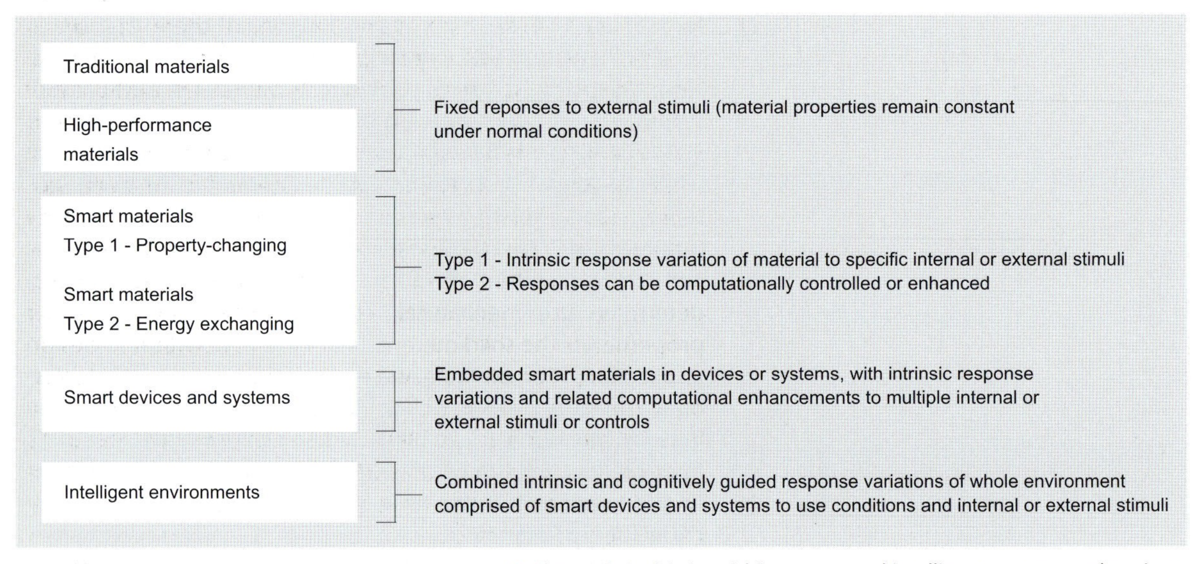

▲ **Figure 2-4** Distinguishing smart and intelligent systems and environments

classification approach based on function/system overlay is shown in Figure 2–4. The figure describes a proposed organization that establishes a sequential relationship between materials, technologies and environments. Cognizant of the need for contextualization, this organization also maintains the fundamental application focus of the more traditional classification system. We will see later that this approach presents other difficulties, but it nevertheless provides a useful way of approaching the subject. The organization of this book, then, mirrors the organization of our proposed classification system.

2.4 The internal structure of materials

Regardless of the classification system used, designers must be exposed to the essential determinants of material behavior. Knowledge of atomic and molecular structure is essential to understanding the intrinsic properties of any material, and particularly so for smart materials. In this section, we begin by briefly reviewing several important topics essential to this understanding. We will see that there are various ways solid materials are composed into the major categories of crystalline solids, amorphous solids and polycrystalline solids. For example, crystalline solids have an orderly and repetitive arrangement of atoms and molecules held together with different types of chemical bonding forces. These patterns form regular lattice structures, of which there are many different types with corresponding material structures. Amorphous solids have a random structure, with little if any order to them, and also have little intrinsic form. Polycrystalline solids are composed of large numbers of small crystals or grains that are in themselves regular, but these crystals or grains are not arranged in any orderly fashion. The precise makeup of these different internal structures and the bonding forces between them largely determine the mechanical, electrical, chemical and other properties of the solid material that are so important in design applications. For example, we have seen earlier that the 'color' of a material depends both on external factors (e.g., the wavelengths of the incident light) and on the material's internal absorption characteristics, which in turn are dependent on the specific organization of the atomic structures that comprise the material.

In order to understand how these different internal structures ultimately determine the resultant properties of materials, it is useful to first look at the different kinds of bonding forces that exist between collections of atoms that ultimately comprise the basic building blocks of any material. Subsequently, the ways individual atoms aggregate into crystalline, amorphous or polycrystalline structures will be reviewed.

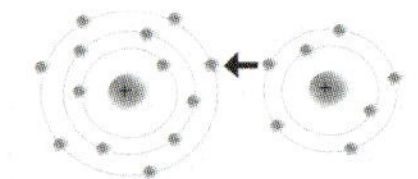

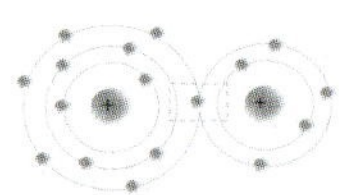

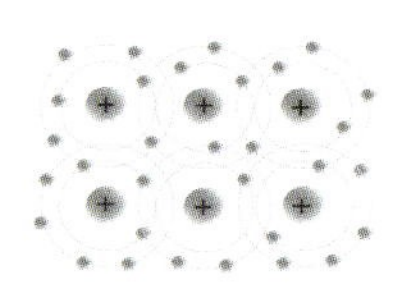

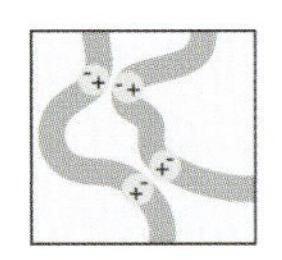

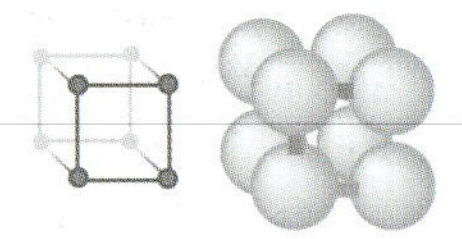

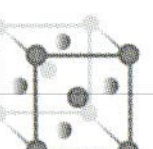

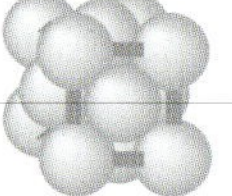

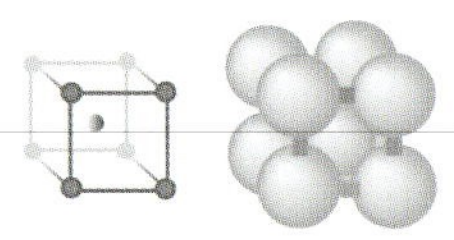

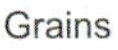

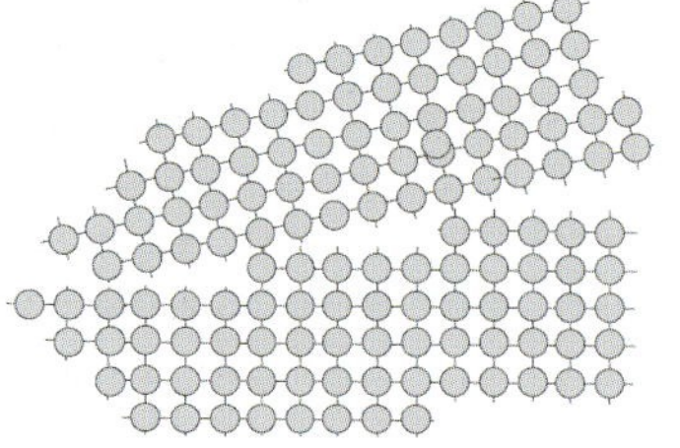

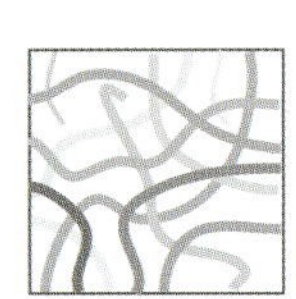

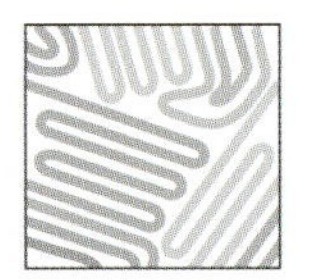

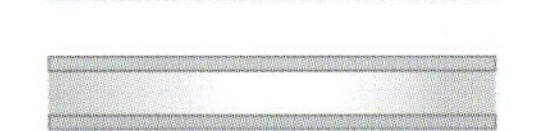

▲ **Figure 2-5** General structures of materials at the micro and macro levels. The structure of a material at each of these levels will strongly influence the final characteristics and properties of the material

BONDING FORCES

At the most fundamental level, we know that an atom consists of three subatomic particles – electrons, protons and neutrons. Electrons revolve at different distances around a positively charged nucleus formed of protons and neutrons. The negatively charged electrons exist at different energy levels in 'shells'. While most of the mass of an atom is concentrated in the nucleus, the nature of the electron cloud is the most significant determinant of the resulting properties. The electrons in the outermost shell are the valence electrons and these are the ones that can be gained or lost during a chemical reaction.

Some atoms do have stable electron arrangements and can exist as single atoms – these are the noble gases. More typically, however, atoms tend to bond to one another to become electronically more stable, consequently forming crystals and molecules. Bonding forces that develop among constituent atoms or molecules hold these larger structures together.

The three primary atomic bonds that develop among atoms are ionic bonds, covalent bonds and metallic bonds. Ionic bonding involves one atom transferring electrons to another atom, covalent bonding involves localized electron sharing and metallic bonding involves decentralized electron sharing. Some secondary atomic and molecular bonds also exist, of which the Van der Waals forces are of primary interest.

In ionic bonding, one atom transfers electrons to another atom to form charged ions. The atom that loses an electron forms a positive ion (electropositive) and is normally considered a metallic element. The one that gains an electron forms a negative ion (electronegative) and is normally considered a non-metallic element. Oppositely charged ions attract one another. The forces associated with ionic bonding thus involve the direct attraction between ions of opposite charge.

Multiple ions typically form into compounds composed of crystalline or orderly lattice-like arrangements that are held together by large interatomic forces. The positive and negative ions form into specific structures whose geometry depends on the elements bonded (crystalline structures are discussed in more detail below). Common table salt or sodium chloride ($NaCl$) has an ionic bond, as do metal oxides. Ionic compounds are solid at room temperatures, and their strong bonding force makes the material hard and brittle. In the solid state, all electrons are bonded and not free to move, hence ionic solids are not electrically conductive (electricity is normally carried by freely moveable electrons). Solid materials based on ionic bonding have high melting points and are generally transparent. Many are soluble in water. In the melted or dissolved state, electrical conduction is possible because both states involve conditions that free up electron bonds and make them moveable.

When atoms locally share electrons, covalent bonds are produced. For example, two atoms share one or more pairs of electrons. Unlike ionic bonding, neither atom completely loses or gains electrons. There is a mutual electrical attraction between the positive nuclei of the atoms and the shared electrons between them. This kind of bonding frequently forms between two non-metallic elements. The bonds occur locally between neighboring atoms thereby producing localized directions. In some instances, small covalent arrangements of atoms or molecules can be formed in which individual molecules are relatively strong, but forces between these molecules are weak. Consequently, these arrangements have low melting points and can weaken with increasing heating. They are also poor conductors of electricity. In other instances, such as carbon in the form of diamond, it is possible for many atoms to form a large and complex covalent structure that is extremely strong. These covalent solids form crystals that can be thought of as a single huge molecule made up of many covalent bonds. In diamond, for example, each carbon atom is covalently bound to four other carbon atoms in a tetrahedronal fashion. Covalent structures of this type of structure are normally very hard, have very high melting points, will not dissolve in liquids and, because electrons are closely bound and not free to move easily, they are typically poor electrical conductors.

Metallic bonds are closely related to covalent bonds in that electrons are again shared, but this time in a non-localized, non-directional nature. These kinds of bonds exist in metals such as copper. A characteristic of a metallic element is that it contains few electrons in the outer shells (either one, two or three). Their outer shells are thus far from full. Immediate sharing with localized neighbors will not be able to fill this shell. Rather, electrons in the valence shell are shared by many atoms instead of just two. These shared electrons are not tightly bound to any one atom and move freely about. The forces of attraction between these mobile electrons and the positive metal ions hold the material together. These forces are known as metallic bonds. As a consequence of these forces, the ions tend to arrange themselves in close-packed orderly patterns. These kinds of metals conduct electricity well because of the freely moving electrons. (If a voltage is applied, the electrons move readily – electrons can enter the arrangement and force others out, yielding a current flow). These same arrangements are also good heat conductors, again because of the free electrons. As will be seen later, the same arrangements often allow the material to deform in a ductile way.

A final bonding force to be considered – the Van der Waals bond – occurs between individual molecules. In many materials, particularly polymers, individual molecules are made of covalently bonded atoms and are consequently quite strong. Due to the normal imbalance of electrical charges between molecules, small attractive forces – the Van der Waals bonds – are developed between them. These secondary bonding forces are relatively weak by comparison to ionic, covalent and metallic bonds. They can break easily under stress and they allow molecules to slide with respect to one another. Ice crystals, for example, are strong H_2O molecules bonded to one another by Van der Waals bonds, but heat or pressure causes these bonds to break down, resulting in liquid phase water.

In summary, the atomic bonding forces determine many of the properties of the final material. These forces are by no means equally strong. In general terms, ionic bonding is the strongest, followed by covalent bonding, metallic bonding and, finally, Van der Waals bonds, which are the weakest of all. While defining material types solely by bonding forces alone is not adequate, we can none the less still observe the following: (1) metals are materials characterized by metallic bonds; (2) ceramics are polycrystalline materials based on ionic and/or covalent bonds; (3) polymers have molecular structures (chains of atoms) that are covalently bonded, but with the chains held together by Van der Waals forces. Further differentiations will be discussed below.

CRYSTALLINE SOLIDS, AMORPHOUS SOLIDS AND POLYCRYSTALLINE SOLIDS

The physical structure of materials is characterized by the arrangement of their atoms, ions and molecules. In the discussion above, it was noted that individual atoms typically bond to one another to become electrically stable and form larger structures. The characteristics of the individual atoms that are bonding, and the kind of bonding force that exists among them, largely determine the way that they aggregate. For example, it was noted that in diamond each carbon atom is covalently bonded with four others to produce a tetrahedronal arrangement. This basic arrangement can be repeated many times over to create a large crystalline structure.

Atomic arrangements in a crystal are described by the spatial network of lines defined by the location of atoms at the intersection points. The idea of a unit cell that specifies atom positions is used as the conceptual building block of a crystal since it forms a basic repetitive unit. The characteristics and geometry of the unit cell are determined by its basic atomic structure. A crystalline structure is made up of large number of identical unit cells that are stacked together in a repeated array or lattice. The shape or geometry of the resulting crystal depends in turn on that of its constituent unit cells. A close study of the geometry of unit cells reveals that there are really only seven possible basic types: cubic, tetragonal, orthorhombic, rhombohedral, hexagonal, monoclinic and triclinic. These basic cells can then be replicated to form identifiable larger lattice structures. Basic morphological considerations indicate that there are 14 basic lattice structures (known as Bravais Lattices) that can be made from the seven basic unit cells (some basic unit cell types can repeat themselves in multiple ways). For example, one of the basic lattices is called a face-centered cubic lattice. In this lattice, atoms are located at the eight corners and the centers of the six faces. Copper, for example, has a face-centered cubic lattice. By contrast, in a body-centered structure there is a single atom at the center of each unit cell with others at the corners or sides of the cell. Tungsten, for example, is a body-center cubic structure, as is iron. Other lattices have different arrangements that in turn can be identified with different real materials. Many typical metals, for example, have either a face or body-centered cubic structure, or a close-packed hexagonal one. Titanium, for example, has a hexagonal close-packed structure, as does zinc.

A particular crystalline structure can become quite large in physical terms. For a number of reasons, however, the growth of a crystalline pattern is interrupted and a grain is formed. A grain is nothing more than a crystalline structure without smooth faces. Many materials are composed of large numbers of these grains. Particular grains meet one another at irregular grain boundaries and are normally randomly oriented to one another. Grain size can vary due to multiple reasons. Metal-working procedures – including heat treatment, cold working or hammering – alter grain size and orientation (changes are visible in a microscope). Alterations in the grain structure can in turn produce changes in material properties (e.g., ductility, hardness).

In a more general sense, it is important to note that material properties are affected not only by the type of crystalline structure present and the macro-structural properties such as grain arrangement, but also by other factors. It is extremely important to note that a pure crystalline arrangement can have enormous strength. Based on studies of bonding strengths and lattice arrangements, material scientists can calculate so-called 'theoretical strengths'. Actual tests of very small ideal specimens (often called 'whiskers') reveal that actual strengths can match theoretical strengths. Early tests on tiny tin crystals demonstrated strengths of over a million pounds per square inch. Even tiny glass fibers, for example, demonstrated tensile strengths of up to 500 000 psi (3450 Mpa) – a value that is about six times higher than that of high-strength steel.

Tests on larger specimens, however, suggest that these maximum strengths cannot normally be obtained. This is because of the normal and expected existence of micro-defects in lattice structures. These include point defects (missing atoms), line defects (a row of missing atoms), area defects (including grain boundaries previously noted) and volume defects (actual cavities). All of these variations from the perfect lattice typically cause changes in the properties of materials, particularly metals.

Line defects, typically called 'dislocations', are particularly important in understanding the differences between theoretical strengths and actual strengths. A missing line of atoms might cause a line defect, or the inclusion of an extra line that in turn causes an opening in the crystalline structure. Under the application of a stress, these dislocations actually move through the structure of the crystal.

Materials in which dislocation movements freely occur are normally very ductile (i.e., they deform plastically very easily). Typical processes of rolling, casting and subsequent heat or mechanical treatments of larger material pieces can create literally millions of dislocations in a crystalline structure. These same processes also affect grain size and other characteristics. Together, the properties (e.g., strength, ductility, malleability) of many common metals are strongly influenced.

Other materials cannot be similarly characterized. As will be discussed more below, many polymeric materials are long chain molecular structures. The individual polymeric chains themselves are normally covalently bonded and quite strong. The connections, however, from chain to chain are held together by weaker Van der Waals bonds. Long chain molecular structures can be cross-linked or folded, which in turn gives the final material different characteristics and properties.

PHASES

Many of the materials that we are interested in have multiple phases, of which the major ones are gas, liquid and solid. We are familiar with the phase transformations as ice melts or as water vapor condenses. These phase changes occur because of changes in their temperature or surrounding pressure. Besides these major phases, many materials have transition states that produce incremental phase changes as the material undergoes a change in its environmental conditions. Iron, for example, has a particular crystalline structure (BBC or body centered cubic) at room temperature. Its phase in this normal state is called ferrite. Upon heating to above 1644 °F the internal structure of iron changes to a new crystalline form (FFC or face-centered cubic) that is called austenite. It undergoes yet another phase change at around 2550 °F since above this temperature austenite is no longer a stable form of iron. This final phase just before melting temporarily reverts back to a body-centered cubic structure. Mechanical and other properties change as the material undergoes each of these phase transformations.

Phase diagrams are used graphically to represent what phases or states exist in a material at different combinations of temperature, pressure and composition. In a pure element, such as iron, composition is not a variable. In alloys, including steel, the precise composition of metals making up the alloy is a critical factor in determining how the mixture varies under different temperature and pressure states. Phase diagrams require some experience in learning how to read them, but they are a common tool of the chemist and materials scientist in understanding the behavior of materials under different conditions.

2.5 Properties of materials

Materials are often distinguished by their properties, some of which are intrinsic and some of which are extrinsic. An intrinsic property is determined by the molecular structure – essentially the chemical composition – of the material. As such, the definition of a specific material also defines all of its intrinsic properties. For example, strength is related to the interatomic forces within the molecule in conjunction with the intermolecular forces that form the material structure: the higher the forces, the greater the strength and hardness of the material. These same forces also directly correlate with the substance's melting and boiling points. A material like diamond, with strong interatomic and intermolecular forces, is not only one of the hardest materials in existence, but also has an extraordinarily high melting point. Besides strength, commonly recognized as intrinsic are a material's mechanical properties, including elastic modulus, and toughness, its physical properties, including conductivity, specific heat and density, and its chemical properties, including reactivity, valence and solubility.

Extrinsic properties are those defined by the macro-structure of the material and as such cannot be directly determined by the composition alone. The optical properties of a material – reflectivity, transmissivity, absorptivity – are often extrinsic as are many acoustic properties. Simply polishing the surface of a metal will produce a substantial change in its reflectivity. Several extrinsic properties are also dependent upon the characteristics of the energy fields of their environment. The color of a material is not a property of the material *per se*, as it is completely dependent on the spectral distribution of the incident light.

Property changes, then, can be produced either by an alteration of the composition of the material or by an alteration in the micro-structure of the material. Both alterations are initiated by an energy input to the material. Input energy can be in many forms of which the most common for smart materials include electrical, chemical, thermal, mechanical and radiative. While most materials undergo similar property changes with an input of energy – for example hot rolling steel changes its microstructure and therefore its properties – smart material changes are also reversible: when the energy input is removed, the material reverts back to its original properties.

All material properties, whether intrinsic or extrinsic, smart or 'dumb', fall into one or more of five categories. The categories – mechanical, thermal, electrical, chemical and optical – are indicative of the energy stimuli that every material must respond to. Although we will study energy stimuli in depth in Chapter 3, a few basics now will be helpful in developing a qualitative understanding of the deterministic relationship between a material and its properties. All energy stimuli are the result of 'difference'. A difference in temperature produces heat, a difference in pressure produces mechanical work. Properties are what mediate that difference. As such, we will note that properties generally have units that reflect the nature of the difference.

Mechanical properties determine how a material will behave when subjected to a load or a mechanical force. That load may be produced by a weight, a shear force, impact, torsion, or a moment, and the behavior that results from these loads includes strain, deformation, or fracture. The mechanical properties determine what result will be produced, and to what degree, by the application of a specific load. Mechanical properties depend on the type of interatomic bonds present, the arrangement of atoms, their larger-scale organizations, the presence of dislocation mechanisms, and gross physical characteristics such as grain size and boundaries. These factors are in turn influenced by material type and composition, including alloying, and if the material is subsequently treated or processed (e.g., annealing, tempering or work-hardening for metals). Mechanical properties are described by specific measures. Strength is a measure of a material's resistance to forces and is commonly described in terms of failure or yield stress (force per unit area) levels. Strain is a deformation measure. Stiffness is a measure of the stress-deformation characteristics of a material. For materials with linear stress-strain characteristics, the Modulus of Elasticity is a useful descriptor of the stiffness characteristics of a material. There are different failure stress levels and elastic moduli depending on the state of stress (tension, shear). Additional mechanical properties of interest include a material's ductility or brittleness, malleability, toughness, hardness, fatigue limits, creep characteristics and others. These properties are discussed at length in many other books.[2]

Electrical properties of primary interest include a material's conductivity and resistivity. Conductivity – the ability of a material to conduct electrical current – is so important that materials are often classified by this property into conductors, semiconductors and nonconductors (insulators or dielectrics). Resistivity is the inverse of conductivity. Materials with a lot of free electrons (e.g., metallic bonded materials) are good conductors since the free electrons become carriers of electrical current. In general, the conductivity of a material increases with lowering temperatures. Superconductivity refers to a phenomenon in materials below a certain critical temperature where resistivity almost vanishes. Special semiconductor materials, however, can behave differently wherein conductivity can increase with rising temperatures (see Chapter 4). Magnetic properties are closely allied to electrical phenomena and properties. Computer disks, motors and generators, credit cards, etc., all are based on magnetic phenomena. Depending on how the material responds to a field, magnetic materials are classified as ferromagnetic, antiferromagnetic, ferrimagnetic, diamagnetic and paramagnetic.

Thermal properties of fundamental significance include a material's thermal conductivity, heat capacity and thermal expansion. Thermal conductivity in conductive materials, such as metals, can be largely explained in terms of the free electron movements discussed earlier. Thermal energy in the form of rapid atomic lattice movements is transmitted through a material via electron movements from the hot to the cold end. Thermal conductivity in dielectric (insulating) materials is a more complex action as it occurs through vibratory phenomena since few or no free electrons exist. Heat capacity is a measure of the amount of heat needed to be transferred to a material in order to raise its temperature a certain amount. Thermal expansion refers to the amount of dimensional change that occurs in a material as a consequence of a temperature change. Most materials, with the notable exception of water changing to ice, tend to shrink with decreasing temperature levels.

Chemical properties of particular interest include a material's reactivity, valence and solubility. Reactivity is a measure of how a material chemically acts with another substance to produce a chemical change. The term solubility generally refers to the capability of a material for being dissolved (a solvent, in turn, is a material, usually liquid, that has the capability of dissolving another substance). The term valence generally refers to the capacity of an element to combine with another to form molecules.

The optical properties of a material, such as its reflectivity, transmissivity and absorptivity, are complex since they may depend upon both intrinsic and extrinsic factors. When light is incident on a material, it is either re-emitted via reflection or transmission, or it might be converted into heat energy. These phenomena are closely dependent on the nature of the material's electron field at its surface (see Chapter 4).

2.6 General classes of materials

Briefly, there are three primary material classes – metals, ceramics, polymers – and many related or derivative types fall into a fourth type known as composites.

Pure metals, including copper, are characterized by their metallic bonds and regular lattice structures. Many metals having face-centered cubic organizations are quite ductile because external forces easily cause slipping among planes that have preferred directions. Iron and nickel are transitional metals involving both metallic and covalent bonds, and tend to be less ductile than other metals. Dislocations and related phenomena are of extreme importance in understanding how metals behave.

Ceramics are characterized by their strong ionic and covalent bonds. Since there are no free electrons that move around, these materials have crystal structures that are electrically neutral and are not good conductors. Dislocation movements are present in ceramics but are of lesser importance. In general, ceramics are hard and brittle, and tend to fail along special cleavage planes. Consequently, ceramics are normally very hard and brittle. They have high resistance to heating and are often used as refractory materials. Glass is an amorphous non-crystalline structure linked by covalent bonds.

Polymers are composed of long-chain molecular structures. Individual molecules are covalently bonded. In simple *thermoplastics*, the chains are not directly connected but are bound together only by weak van der Waals bonds. Hence, they are quite soft and ductile since external forces can cause chains to slide by one another relatively easily. These same thermoplastics can be easily melted (heat breaks down the van der Waals bonds) and will then reform into a solid when cooled. Thus, they can be easily recycled. *Thermosets*, by contrast, have additional hardeners added to them that cause the long-chain molecules to be cross-linked or interconnected. Common epoxies are thermosets. External forces cannot cause chains to easily slide by one another. Consequently, these materials can be quite strong and hard. They cannot, however, be melted like thermoplastics.

Folded chain polymers have a periodic arrangement of chains in them that are crystalline in nature, but not cross-linked, and have multi-layered structures to them. They can be formed in many ways, including crystallization from dilute solutions. These semi-crystalline polymers can be quite dense. They can be made chemically resistant and highly heat resistant. In Chapter 4 we will see that this class of polymer is particularly important *vis-à-vis* smart behaviors since the semi-crystalline nature of these folded chains allows many properties to be imparted to them that are not normally associated with polymers (e.g., conductivity).

Elastomers are polymers that have largely amorphous structures, but are lightly cross-linked. They can be thought of as laying between thermoplastics and thermosets. Many natural materials are elastomeric whereas other elastomers can be readily synthesized. The Vulcanization process – used in making common automobile tires – creates cross-links containing sulfur atoms. The rubber gives the tires elasticity, but the cross-linking makes them sufficiently stiff and hard.

Composites are high-performance materials that are made by combining two or more primary materials. They comprise

REINFORCING

Reinforcing materials
- Glass fibers
- Polymer fibers
  - Organic (e.g., Kevlar)
  - Nylons, polyesters, etc.
- Carbon fibers

Organization of reinforcing
- Basic forms
  - Strands, filaments, fibers, yarns (twisted strands), rovings (bundled strands)
- Weaves, braids, knits, other
- Nonwoven mattings
- Films, sheets
- Other

Weaves
Knits
Braids
Other

RESINS AND MATRIX MATERIALS

Resin materials
- Epoxies
- Polyesters
- Vinyls
- Other

CORES

Core materials
- Foams
- Balsa
- Synthetic fabrics
- Other

Organization of cores
- Honeycombs
- Laminates
- Other

▲ **Figure 2-6** General makeup of composite materials intended for high performance strength or stiffness applications

Reinforced polymer Dacron/Kevlar fabric

▲ **Figure 2-7** Two flexible composite sheets

a huge class of materials – there are literally thousands of them – and are beyond the scope of this treatment to discuss in detail. Briefly, composite materials are invariably intended for high-performance applications where their properties are engineered for specific purposes, and they may be broadly thought about in terms of their functions. Are they intended to serve strength or stiffness functions? Are they meant to reduce thermal conductivity? Are they meant to have special reflective characteristics? Figure 2–6 shows the general makeup of composites intended for strength or stiffness applications. Normally, these composites are made up of reinforcing materials, resins or matrix materials that the reinforcing materials are embedded into, and, quite frequently, internal cores are present as well. Different forms of these kinds of composites can be engineered for specific strength or stiffness applications, including directions of stresses and so forth. For other purposes, embedded materials

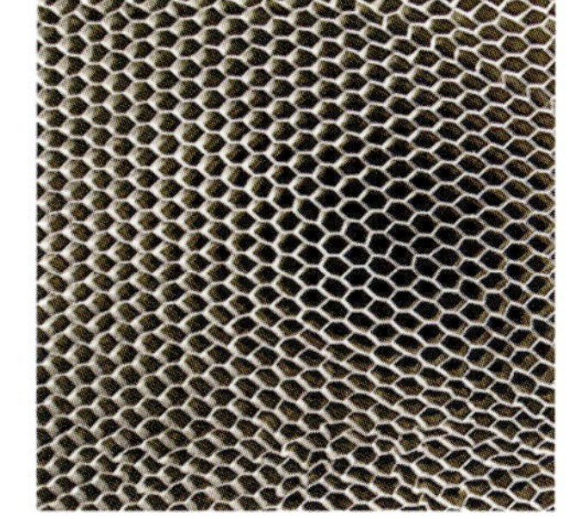

Carbon fiber sheet Aluminum honeycomb core

▲ **Figure 2-8** Typical materials used in composites

may not serve strength functions at all. Fiber-optic cables, for example, have been embedded in different materials to serve as strain or crack detectors in the primary material. Also, different films or sheet products may be laminated together as well. The high performance radiant color films with multiple reflectance qualities, for example, are multi-layered laminates of different types of films.

2.7 Nanomaterials

The term 'nanotechnology' has attracted considerable scientific and public attention over the past few years. The prefix 'nano' indicates that the dimensional scale of a thing or a behavior is on the order of a few billionths of a meter and it covers a territory as large, if not larger, than that represented by micro-scale. For comparison, the head of a pin is about one million nanometers across whereas a DNA molecule is about 2.5 nanometers wide. Given that individual atoms are nanometer size (for example, 5 silicon atoms is equivalent to one nanometer), then the ability to build structures one atom at a time has been a provocative objective for many materials scientists. In its simplest form, nanotechnology conceptually offers the potential to build 'bottom up,'

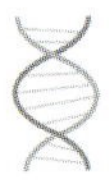
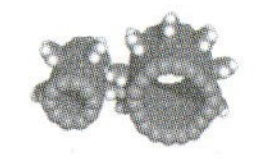

▲ **Figure 2-9** Relative size comparisons. Nano-scale objects exist at sizes near the atomic level. Micro-scale objects, such as many MEMs devices, are much larger (a human hair, for example, is about 50 μm in diameter) and can be visually seen. Devices at the meso-scale level (equivalent to millimeter to centimeter scales) are relatively large in comparison to microscale and nanoscale objects, but still very small with respect to human dimensions creating materials and structures with no defects and with novel properties.

As discussed earlier in this chapter, the precise makeup of different internal structures and the bonding forces between them largely determine the mechanical, electrical, chemical and other properties of the solid material. Nanotechnology, by enabling the complete construction of the molecular structure, may afford us the possibility to design unprecedented and dramatically enhanced properties for the macro-scale. Indeed, it may even be possible to produce substantially different properties without even changing the chemical composition. Already, one nanomaterial – carbon nanotubes – has been attributed with an electrical conductivity that is 6 *orders of magnitude* higher than copper, and a strength to weight ratio that is 500 times greater than that of aluminum.[3] Essentially, we will be able to *program* material properties. Furthermore, constructing bottom up could also allow for self-assembly, in which the random (non-continuum) motion of atoms will result in their combination, or for self-replication, in which growth occurs through exponential doubling.

Beyond the opportunity to 'build' materials from scratch, nanotechnology also encompasses the development of and application of nano-sized materials and systems. Nano-particles are being proposed for inclusion in paints and abrasives, and nanoprobes are intended to be the basis of *in vivo* drug delivery devices. Quantum dots – nanometer-sized semiconductor crystals capable of confining a single electron – represent the next generation in luminescent technology as they essentially are quantum LEDs (light-emitting diodes). The potential applications for nanotechnology abound, from data storage to body armor, but this exciting field is still in its infancy, and many of proposed application domains are, at best, speculative. Nevertheless, both the technologies and ideas implicit in thinking about behaviors at the nano-scale hold great promise for the future.

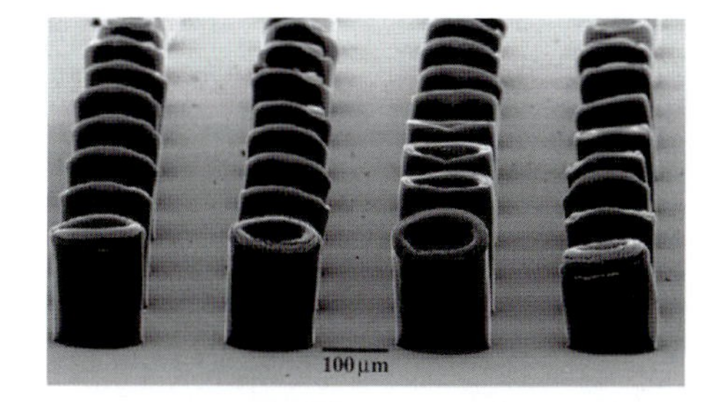

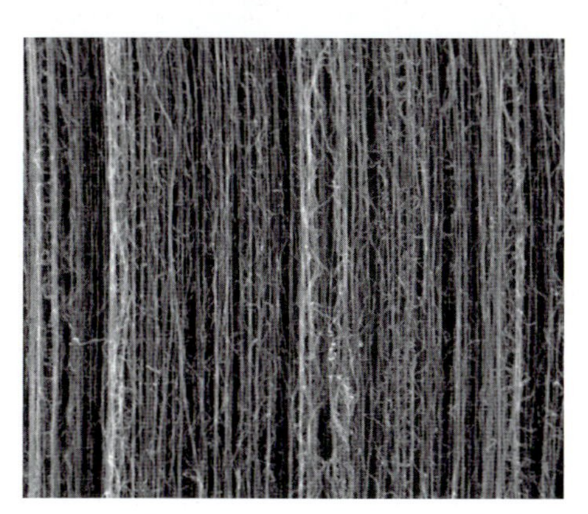

▲ **Figure 2-10** Carbon nanotubes (CNT). A tubular form of carbon, with a diameter as small as 1 nm, is produced from sheets rolled into tubes. (NASA Ames)

Notes and references

1 Braddock, S. and Mahoney, M. (1998) *Technotextiles*. London: Thames and Hudson.

2 See, for example, Schodek, D. (2003) *Structures*, 5th edn. Englewood Cliffs, NJ: Prentice-Hall.

3 These numbers came from a presentation titled 'An Overview of Recent Advancements in Nanotechnology', delivered by M. Meyyappan of NASA Ames Research Center in October 2002. The numbers vary widely from source to source.

3 Energy: behavior, phenomena and environments

Although we often imagine materials as things that can be weighed, measured and described, and thus as tangible artifacts, our primary interest as designers is in how materials behave. A steel column becomes useful when it supports a load. A pane of glass is meaningful when it transmits light. When we choose a material, we inherently choose it for its interaction with some type of energy stimulus, and this is true even for those materials that we simply wish to view, such as those in a sculpture. As a result, no discussion of materials can be complete without an understanding of energy.

3.1 Fundamentals of energy

What is energy? This is a difficult question, as energy is not a material thing at all, even though it is the fundamental determinant of all processes that take place among and between all entities. Whenever an entity – from an atom to an ecosystem – undergoes any kind of change, energy must transfer from one place to another and/or change form. For example, heat must be transferred – added or removed – for the temperature of an object to change, and the form of energy must change from kinetic to electrical when a turbine rotates a shaft in a generator. Conceptually, all forms of energy can be divided into two generic classes:

- Potential energy: stored energy that *can flow*. The energy that is stored by virtue of position, bending, stretching, compression, chemical combination. Examples include water behind a dam, boulder on top of a hill, a coiled spring, gasoline.
- Kinetic energy: energy that *is flowing*. The energy of motion that moves from high potential entities to low potential entities. Examples include water falling over a dam, a boulder rolling down a hill, the combustion of gasoline in an engine.

Within these two classes, energy takes many different forms, and each form is characterized by a fundamental variable that becomes useful as energy only when difference comes into play. Without a difference between one state and another, energy cannot flow. As a result, energy can only be quantified and measured as it moves or by its potential to move.

HEAT	driven by temperature difference
WORK	driven by pressure difference
POTENTIAL:	driven by height difference
ELECTRICAL:	driven by charge difference
KINETIC:	driven by momentum difference
ELASTIC:	driven by deflection difference
CHEMICAL:	driven by atomic attraction differences
NUCLEAR:	driven by quanta differences
MAGNETIC:	driven by moving charge differences

What, then, governs this movement? Although the flow of energy determines the behavior and state of all things – living and inanimate – scientists did not develop an understanding of it until 1850, two centuries after the establishment of Newtonian physics. The 19th century developments in steam engine technology finally led to the discovery of an important principle: the conservation of energy. This principle is perhaps the most fundamental building block of physics, and it is also the foundation for the branch of physics known as thermodynamics – the science of energy.

3.2 Laws of thermodynamics

Derived from the Greek words *thermé* (heat) and *dynamis* (force), thermodynamics describes the branch of physics concerned with the conditions of material systems and the causes of any changes in those conditions. A material system may be comprised of anything from a solid to a liquid to a gas as well as a mixture thereof, but it is distinguishable as an identifiable quantity of matter that can be separated from everything else – the surroundings – by a well-defined boundary. The conditions of a system at any given moment are known as its *state*, which basically encompasses all that can be measured – temperature, pressure and density – as well as its internal energy.

The nature of the relationship between a system and its surroundings is governed by the Laws of Thermodynamics. There are four laws, of which the first three are most relevant for our discussions. Even though each law makes a reference to heat (hence *thermo*-dynamics), together they govern the dynamics of all forms of energy.

ZEROTH LAW OF THERMODYNAMICS (ALSO KNOWN AS 'THE LAW OF THE THERMOMETER')

If two entities are in thermal equilibrium with a third entity (such as a thermometer), then they are in thermal equilibrium with each other. Essentially, this law tells us that equilibrium is a condition without difference, and thus without further energy exchange.

FIRST LAW OF THERMODYNAMICS (ALSO KNOWN AS 'THE LAW OF CONSERVATION OF ENERGY')

While energy assumes many forms, the total quantity of energy cannot change. As energy disappears in one form, it must appear simultaneously in other forms – energy is thus indestructible and uncreatable (in the Newtonian world-view). More formally, the rate of energy transfer into a system is equal to the rate of energy transfer out of a system plus any change of energy inside the system. The First Law can be conceptually represented by the following expression:

$$\Delta \text{ (energy of system)} + \Delta \text{ (energy of surroundings)} = 0$$

If energy is convertible and indestructible, then it must be possible to measure all forms of it in the same units. Regardless of whether the energy is electrical, or thermal, or kinetic, we can measure it in kilowatt-hours, and convert it into calories, BTUs, foot-pounds, joules, electron volts and so on. While it may be difficult to imagine that one could talk about foot-pounds of heat, or calories of electric current, the First Law establishes their equivalence.

The generation of electricity in a power plant is an excellent example of the First Law, as energy must go through many transformations before it can become directly useful at a human scale. The combustion of coal (chemical energy) produces the heat that converts water into steam (thermal energy) that is used to drive a turbine (mechanical energy) that is used to rotate a shaft in a generator thereby producing electrical energy. These are just the energy exchanges within a power plant, we could also extend the transformations in both directions: the chemical energy in the coal results from the decay of plant materials (more chemical energy) which originally received their energy from the sun (radiant energy) where the energy is produced by fusion (nuclear energy), and so on. In the other direction, electricity produced by the power plant might be used to run the compressor (kinetic energy) of a chiller that provides chilled water (thermal energy) for cooling a building.

This tidy accounting of energy might lead one to conclude that there cannot be a global energy problem, as energy is never destroyed. This, however, is where the Second Law comes into play.

SECOND LAW OF THERMODYNAMICS (ALSO KNOWN AS 'THE LAW OF ENTROPY' OR 'THE CLAUSIUS INEQUALITY'

Entropy is an extensive property of a system that describes the microscopic disorder of that system. Whenever a process occurs, the entropy of all systems must either increase or, if the process is reversible, remain constant. In 1850, Rudolf Clausius stated this in terms of directionality: 'It is impossible to construct a machine operating in a cyclic manner which is able to convey heat from one reservoir at a lower temperature to one at a higher temperature and produce no other effect on any part of the environment.'[1] In other words, there is a natural direction to processes in the universe, resulting in an energy penalty to move in the opposite direction. Water above a waterfall will naturally flow to a lower level, but it must be pumped up from that level to return to its starting point.

Although the second law is often rhetorically interpreted as 'increasing randomness', entropy is neither random nor chaotic. The concept of 'exergy' explains just what the penalty is when we attempt to reverse a process.

EXERGY (ALSO KNOWN AS AVAILABILITY)

The exergy of a thermodynamic system is a measure of the *useful* work that can be produced in a process. Work is any interaction between a system and its surroundings that can be used to lift a weight, and as such, work is harnessable. *Lost work* is the difference between the ideal work and the work actually done by the process. Basically, even though the laws of thermodynamics state that energy can never be destroyed, *lost work* is that which has been wasted, in the sense that it can become unavailable for further transformation, and thus unavailable for human use. Wasted work turns up as heat. So, for example, if a generator converts kinetic energy into electrical energy at an efficiency of 90%, then 90% of the initial energy produces work, and the remaining 10% produces heat. Referring back to the Second Law, we begin to recognize that, on a universal level, every single process is reducing the amount of concentrated energy available while increasing the amount of distributed (and therefore, unharnessable) heat.

With this understanding of the rules by which energy is converted from one form to another, we can now express the First Law more formally:

$$\Sigma Q \text{ (heat)} - \Sigma W \text{ (work)} = \Delta U \text{ (internal energy)} + \Delta E_k \text{ (kinetic energy)} + \Delta E_p \text{ (potential energy)}$$

Both heat and work are transient phenomena; systems do not possess heat or work as they might possess internal or potential energy. Instead, heat and work are only manifested by the transfer of energy across the boundary between a system and its surroundings. As such, a thermodynamic boundary is a region of change, rather than a discontinuity.

Why is the study of thermodynamics important for understanding the behavior of materials and, more importantly, that of smart materials? For architects, the most typical interaction for a material is the load produced by gravitational forces. As a result, properties represented by Young's modulus or the yield point are the most familiar. Classical discussions of mechanics would suffice. But, as mentioned earlier, the behavior of a material is dependent upon its interaction with an energy stimulus. All energy interactions are governed by the laws of thermodynamics, whether it is the appearance of an object in light or the expansion of a material with heat. Material properties determine many aspects of these interactions. For example, one material property may determine the rate at which energy transfers; another property may determine the final state of the object. A general thermodynamic relationship between a material system and its energy stimulus can be conceptualized by the following:

state of the object or material system × property
= function of energy transfer

As an example, if we look at Fourier's Law, which calculates the rate of heat transfer through a material, we can begin to see how the material property of conductance determines the state of the object.

$$\Delta T\,(U \times A) = \Delta Q$$

T = temperature, Q = heat transfer rate,
U = conductance, A = area

The state of the object (or material system) is denoted by the state variable of temperature, whereas the heat transfer rate represents the amount of energy exchanged or transformed by the object. The area is an indication of how much material is being affected, and the property of conductance ultimately determines either what the temperature of the object will be or how much heat must transfer in order for the object to reach a particular temperature.

We can use this conceptual thermodynamic relationship between a material system and its energy stimulus as a framework for organizing material behavior. In traditional materials, as well as in many high performance materials, properties are constant over the range of state conditions faced in the typical application. For example, the conductance of steel is constant at temperatures from 32 °F to 212 °F (0–100 °C), and only when the temperature reaches approximately 1000 °F (approx. 535 °C) will the drop in conductance no longer be negligible. As such, for a given material in this category, the state of the object is primarily a function of the energy transfer. In Type I smart materials, properties will change with an energy input. For example, the transmittance of electrochromic glazing – in which the molecular properties of a coating are changed by application of a current – can be switched by a factor of ten. In this category, then, the property is a function of the energy transfer. Type II smart materials are energy exchangers, transforming input energy in one form to output energy in another form. A photovoltaic is a common Type II material; through the conditions of its state, input solar radiation is converted into an electrical current output. The property of the material may be instrumental in producing the exchange but it is not the focus of the object's behavior. We can now summarize the three conceptual thermodynamic relationships for each of these categories as follows:

- Traditional material: State of the object = *f* (energy transfer), property = constant.
- Type I smart material: Property = *f* (energy transfer), state of object may change.
- Type II smart material: Energy transfer = *f* (state of the object), property may change.

3.3 The thermodynamic boundary

The further completion of this thermodynamic conceptualization of materials requires that we also understand the concept of boundary as behavior. This is particularly difficult for architects and designers as our more normative definition of boundary directly refers to lines on drawings. Walls, rooms, windows, façades, roofs and property lines depict boundary in the lexicon of design. As discussed in Chapter 1, thermodynamic boundaries are not legible and tangible things, but instead are zones of activity, mostly non-visible. In this zone of activity – the boundary – the truly interesting phenomena take place. This is where energy transfers and exchanges form,

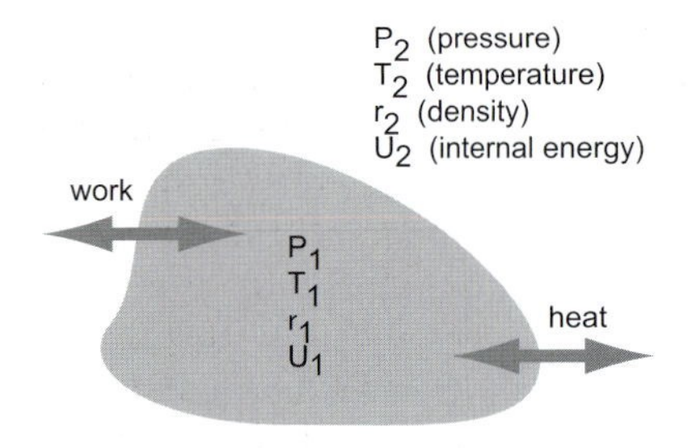

▲ **Figure 3-1** Thermodynamic system. An energy state is any identifiable collection of matter that can be described by a single temperature, pressure, density and internal energy. The boundary differentiates between distinct states. Only work or heat can cross the boundary

and where work acts upon the environment. By examining a simple diagram of a thermodynamic system, we see that the boundary demarcates the *difference* between the material at its identifiable state and the immediate surroundings in a state that may vary in temperature, pressure, density and/or internal energy. While diagrammatically this boundary appears to be a discontinuity or a border, physically it is where the mediated connection between the two states occurs. All change takes place at the boundary.

In most disciplines in which the laws of physics, and particularly those of thermodynamics, are fundamental to the development of the applied technologies, the boundary operates as the fundamental transition zone for mediating the change between two or more state variables. For example, when a warm air mass is adjacent to a cool air mass, such as in a warm front, each of these masses will have a distinguishable temperature and pressure. A boundary layer will develop between these masses, and the transition in temperature and pressure will occur in this layer. This mitigating boundary occurs at all scales, from that of the atmosphere to a microchip, and it is fundamentally responsible for the thermal well-being of the human body.

One of the most common thermodynamic boundaries in a building happens to be located next to the most commonly drawn boundary – that of the wall. The boundary of interest here is not the one we routinely think of – the wall as solid boundary between inside and outside – but rather it is the boundary layer between the wall as a material object and the adjacent air as the surrounding environment. If we compare the two images in Figure 3–3, a number of key differences stand out. The boundary layer surrounding the body has a

▲ **Figure 3-2** Warm front. The boundary between the two pressure systems is clearly demarcated by the cloud layer. (NOAA)

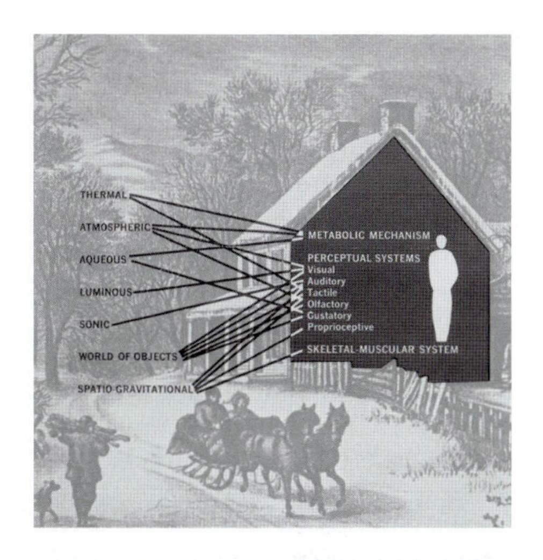

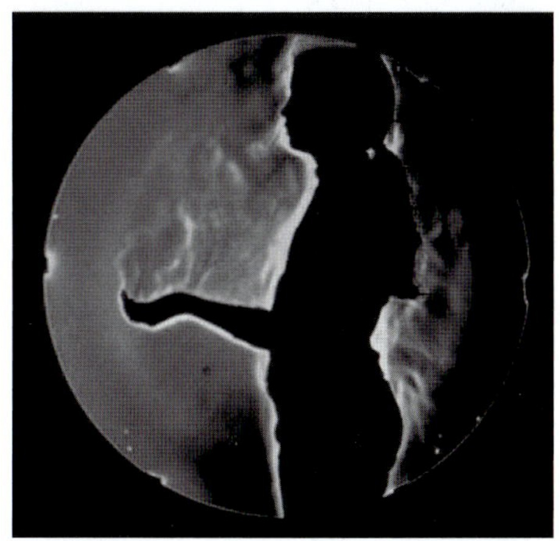

▲ **Figure 3-3** Comparison between architectural depiction of an environmental boundary (top) and that of the physicist (bottom). Image on top is from James Marston Fitch's seminal text *American Building 2: The Environmental Forces that Shape It* (1972). Image on bottom is of convective boundary layer rising from a girl. (Image courtesy of Gary Settles, Penn State University)

non-visible and transient shape, contiguous with the material object, but contingent on the surrounding environment. It only comes into existence if there is a difference in state variables, and its behavior is unique at any given moment and location. In contrast, the building wall exists as an independent element separating two other environments – inside and outside. It does not move, its shape does not change, and most importantly, it does not mediate between the state variables – the continuity of the boundary layer is negated by a discontinuous barrier.

The above example is but one of the many different boundary conditions between material systems and their surrounding environments. Exterior walls also have transient boundary layers. Note in Figure 3–4 how the velocity profile changes in section, even though both the wall and the surrounding environment – the boundary conditions – are stationary. Much more common, and much less identifiable, are boundaries with fluid and moving borders, rather than with one or more solid and stationary borders. We recognize this variation when smoke rises from a burning cigarette or when we release an aerosol from a spray can. This type of boundary condition, termed free field, is ubiquitous and pervasive – every small change in air temperature or pressure will instantaneously produce a mediating boundary that will disappear when equilibrium is reached in that location.

Just as the understanding of thermodynamics helps us to understand the role of materials in an energy field, then this clarification of the boundary can help us to define and create energy environments. In the discipline of architecture, the term environment has typically been used to describe ambient or bulk conditions. The assumption is that the surrounding environment is *de facto* exterior to a building and defined by regional climatic conditions. And the thermodynamic 'material system' has been simplified as the interior of a building with relatively homogeneous conditions. The physics of the building is presumed to be coincident with and defined by the visible artifacts of the building. But while building scale is relevant for many characterizations of architecture, from construction to occupation, it has only a minor relationship with the scale and location of thermodynamic boundaries. When we talk about scale in architecture we often use expressions like macro-scale to represent urban and regional influences and micro-scale to represent building level activities. In contrast, thermodynamic boundaries are often several orders of magnitude smaller. For example, in order to introduce daylight to the interior of the building, architects typically shift the orientation of the façades and

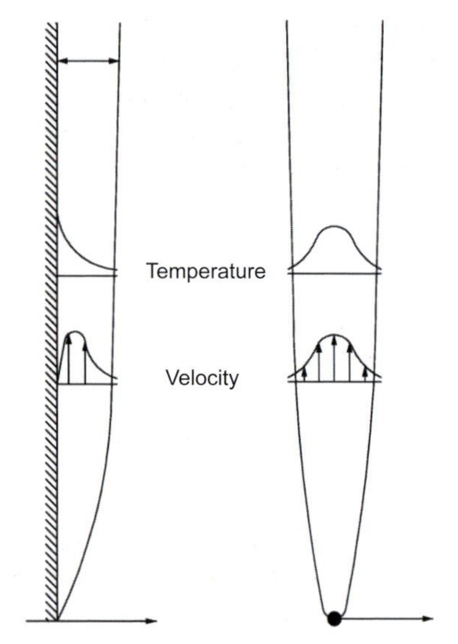

▲ **Figure 3-4** Typical convection behavior in buildings. *Left*, convection against a heated or cooled surface. *Right*, convection above a point source such as a lamp, human or computer

enlarge glazed surfaces. Light, however, is a micron-sized behavior, and the same results can be produced by microscopic changes in surface conditions as those occurring now through large changes in the building. By considering scale in our new definition of boundary as a zone of transition, we can begin to recognize that energy environments – thermal, luminous and acoustic – are rarely 'bounded' by architectural objects. Instead, these energy environments may appear and disappear in multiple locations, and each one will mark a unique and singular state. Our surrounding environment is not as homogeneous as we assume, but rather it is a transient collection of multiple and diverse bounded behaviors.

3.4 Reconceptualizing the human environment

James Marston Fitch, as one of the 20th century's most notable theoreticians of the architectural environment, cemented the concept of architecture as barrier in his seminal book *American Building: The Environmental Forces that Shape It*.

> The ultimate task of architecture is to act in favor of man: to interpose itself between man and the natural environment in which he finds himself, in such a way as to remove the gross environmental load from his shoulders.[2]

The interior is characterized as a singular and stable environment that can be optimized by maintaining ideal conditions. Indeed, one of the most prevalent models of the 'perfect' interior environment is that of the space capsule. The exterior environment is considered fully hostile, and only the creation of a separate and highly controlled interior environment can complete this ideal container for man. This exaltation of the space environment was the culmination of nearly a century of investigation into defining the healthiest thermal conditions for the human body. In the 1920s, with the advent of mechanical environmental systems, standards for interior environments began to be codified for specific applications. School rooms were expected to be maintained at a constant temperature and relative humidity, factories at another set of constant conditions. Over the course of the 20th century, health concerns waned and the standards were tweaked for comfort. Regardless of the intention, the result was a near universal acceptance of stasis and homogeneity.[3]

This characterization of the interior environment is recognizable to us as analogous to a thermal system in which the interior is the material system, the building envelope is the boundary and the exterior is the surroundings. But if we recast the human environment in terms of our earlier discussions of boundary and scale, we realize that the actual material system is the body, the boundary is the body's energy exchange and the surrounding environment is immediately adjacent to the body. The building's environments might be analogous to this system, but it is an analogy of abstraction rather than of reality.

The design of enclosure is not the design of an environment. All environments are energy stimulus fields that may produce heat exchange, the appearance of light, or the reception of sound. Rather than characterizing the entire environment as being represented by a bulk temperature, or a constant lux level of illuminance, we will define the environment only through its energy transactions or exchanges across boundaries, including those of the human body. This approach is consistent with the current understanding of the body's sensory system. Whether thermal, aural, or optical, our body's senses respond not to state conditions – temperature, light level, etc. – but to the rate of change of energy across the boundary. For example, the sensation of cold does not represent an environment at a low temperature, rather it is an indication that the rate of change of thermal energy transfer between the environment and the body is increasing – the temperature of the environment may or may not be one of many possible contributors to this increase. Essentially, the body is sensing itself through its reaction to the surrounding environment, but not sensing the environment. The ubiquitous real world – the world appropriated by sensation – is not at all what it seems.

3.5 The thermal environment

So what is the thermal environment if it is not simply the temperature of our surroundings? Imagine it as a diverse collection of actions. We have already discovered that only heat and work can cross the boundary. This tells us what, but not how. We know that if there is a difference in temperature, then heat will flow from high temperature to low temperature, but that does not tell us any specifics regarding when, how, through which mechanism or in what location. Essentially, we need to know how heat *behaves*. The subset of thermodynamics known as Heat Transfer defines and characterizes the particular thermal behaviors that are constantly in action around us. Even within a room in which the air seems perfectly static and homogeneous, we will be surrounded by a cacophony of thermal behaviors – multiple types of heat transfer, laminar and turbulent flows, temperature/density stratifications, wide-ranging velocities – all occurring simultaneously. The human body's thermal mechanisms may even be more complex that those of the room. Evaporation joins radiant, convective and conductive heat transfer and balances with both internal and external physiological thermoregulation to maintain the body's homeostasis. The transiency of the human state coupled with the large ranges of all the different mechanisms produces a thermal problem that is most probably unique at any given instant. It is for these complex and highly variable conditions that standard building environmental systems are used. The HVAC (heating, ventilating and air conditioning) system emerged over a century ago, and has undergone very little change in the intervening time precisely because of its ability to provide stable and homogeneous conditions within this transient and heterogeneous environment. The heterogeneity of the different thermal behaviors, however, offers unprecedented potential to explore the direct design and control of our thermal environment by addressing each of these behaviors at the appropriate scale and location. A quick overview of heat transfer and fluid mechanics will establish the complex categories of thermal behaviors with the relevant material properties, while exposing the problematic of using a singular response for all of the different types.

MECHANISMS OF HEAT TRANSFER

There are three primary modes of heat transfer. The relevant state variable for each mode will tell us in which direction energy will flow. For example, if the difference between a system and its surroundings is due to temperature, then we know that heat must transfer from high temperature to low temperature. If the difference between a system and its surroundings is due to pressure, then we known that kinetic energy must transfer from high pressure to low pressure. The mode of heat transfer – conduction, convection and radiation – tells us *how* the energy will transfer, i.e. through direct contact or through electromagnetic waves traveling through open space. Each mode of heat transfer will have a predominant material property; it is the material property that determines how *fast* heat will transfer. Ultimately, rate is the most important aspect, particularly for human needs, and it is also the aspect most in control by the designer through appropriate selection of material properties.

The following equations will quickly become quite complex; indeed, we must recall that the science of heat transfer is

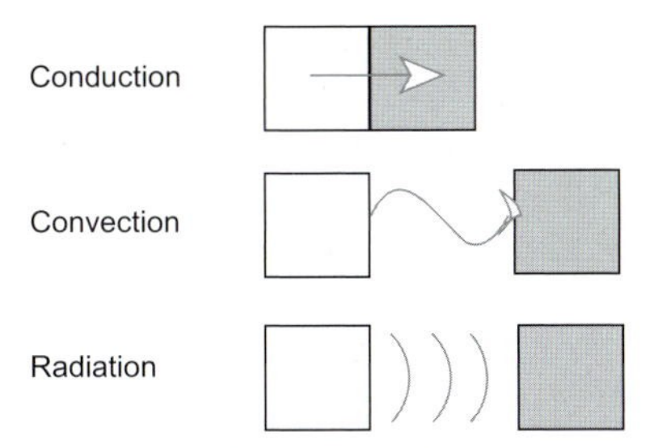

▲ **Figure 3-5** The three modes of heat transfer from a high temperature object to a low temperature object

MATERIAL	CONDUCTIVITY (W/m K)
Copper	406.0
Aluminum	205.0
Steel	50.2
Concrete	1.4
Glass	0.78
Brick	0.72
Water	0.6
Hardwoods	0.16
Fiberglass insulation	0.046
Air	0.024

▲ **Figure 3-6** Thermal conductivities of some typical materials (at 20 °C)

the most difficult, as well as the most recent, branch of classical physics. Nevertheless, we will be able to identify state variables such as temperature and pressure, design variables such as area and thickness, and material properties such as conductivity and emissivity.

(The term 'Heat Transfer' always implies rate, thus all types of heat transfer are in the form of energy change per time change (dQ/dt). Units of Btu/hr or kW are the most commonly used.)

Conduction

Conduction is the mode by which heat is transferred through a solid body or through a fluid at rest. Conduction results from the exchange of kinetic energy between particles or groups of particles at the atomic level. Molecules vibrating at a faster rate bump into and transfer energy to molecules vibrating at a slower rate. In accordance with the Second Law of Thermodynamics, thermal energy transfer by conduction always occurs in the direction of decreasing temperature. Conduction obeys Fourier's Law:

$$dQ/dt = (k/x) \times A \times (T_2 - T_1)$$

where k is the material property of thermal conductivity, x is the shortest distance through the material between T_2 and T_1, and A is the surface area of the material.

The state variable in conduction is temperature, and so we are examining how the difference between these two temperatures is negotiated through a material. Conduction always takes the shortest path possible, so the distance between the two temperatures becomes an important design variable. By increasing the distance (the thickness of the material) x, we can slow down the rate of heat transfer proportionately. For any given thickness, then, the material property of thermal conductivity is the determinant of rate.

Thermal conductivity (k) (units of Btu/ft-hr-°F, kcal/hr-m-°C, W/m-°K) is defined as the constant of proportionality in Fourier's Law. Unfortunately, like many of the terms we use in heat transfer, the definition tends to be described by a process, which, in itself, is described by other processes. As a result, the values of conductivity are determined by experimentation. We can, however, discuss it qualitatively. It is what we call a 'microscopic' property in that it occurs at the atomic level. In metals, the conductivity is due to the motion of free electrons – the greater the motion, the higher the conductivity. In non-metals, or dielectrics, the explanation is more complex: the exchange of energy from atom to atom takes place through 'lattice waves', which is collective vibration as opposed to the individual molecular vibration that we find in the metals. Generally, metals are more conductive than non-metals, and solids are more conductive than liquids and gases in that order.

Convection

Convection is the mode by which heat is transferred as a consequence of the motion of a fluid. Heat can be considered to be transported or 'carried' by the fluid's motion, resulting in the 'mixing' of different energy content fluid streams. The overall process of convection is a macroscopic behavior, but the final temperature change still occurs by the exchange of kinetic energy at the molecular level, just as it does for conduction. *Natural convection* is induced by the natural volume or density changes, coupled with the action of gravity, that are associated with temperature differences in a fluid. *Forced convection* results from fluid motion induced by pressure changes, such as artificially caused by a fan, but can also occur 'naturally' due to the force of winds. In accordance with the Second Law of Thermodynamics, thermal energy transfer by conduction occurs in the direction of decreasing density or pressure. Convection obeys the Navier–Stokes Equations, but solving these three non-linear partial differential equations simultaneously for four variables – pressure (p), temperature (T), density (ρ) and velocity (u) – is one of the most difficult problems in physics and is currently only possible through the use of Computational Fluid Dynamics. Instead, we will focus on the simpler explanations below each of the equations.

Conservation of mass (the continuity equation):

$$\frac{\partial \rho}{\partial t} + \nabla \cdot (\rho u) = 0$$

If we have a volume of a certain size, then the mass flow rate (density × velocity × area) coming in must equal the mass flow rate going out + any change of mass in the volume. Essentially, we must account for all material that enters and leaves, just as we must account for all energy.

Conservation of momentum (Newton's Second Law):

$$\rho \frac{\partial u}{\partial t} + \rho u \cdot \nabla u = -\nabla p + \mu \nabla^2 u + F$$

If we have a volume of a certain size, then the rate of change of its momentum (density × velocity × volume) is equal to the net force acting on it. We often use the more familiar version of this, which is $F = ma$. Any volume that is set into motion, or whose motion changes, will do so in response to either viscous action (fluid friction) or to an external force, such as gravity.

Conservation of energy:

$$\rho C_p \left(\frac{\partial T}{\partial t} + u \cdot \nabla T \right) = \nabla \cdot (\kappa \nabla T) + H$$

In spite of the complexity of this equation, we should recognize it as the First Law of Thermodynamics. If we have a volume of a certain size, the energy coming in must be accounted for by the energy leaving plus any change in the internal energy of the volume.

As we can see, the determination of convection involves many more variables. There are now three state variables – temperature, pressure and density – and as our material system may be set into motion, velocity also becomes a variable. We must also be aware of both interior and exterior factors – wind speed (velocity and pressure), relative location of the temperature difference (density) and the internal energy contained by the fluid (temperature and density). None of these factors came into play in conduction. As in conduction, high temperature will transfer to low temperature, but simultaneously, high pressure will also be transferring to low pressure, and high density will move toward low density – and all of these interact with each other. There are many more design variables – porosity of the building envelope, location and size of openings, the height of surfaces, interior obstructions and building orientation. And joining the thermal conductivity (k) as important material properties are the specific heat (C_p) and the viscosity (μ).

Specific heat (C_p) (units of Btu/lb-°F, cal/g-°C) is defined as the amount of heat required to raise the temperature of a substance or mixture under specified conditions. For example, it takes one calorie to raise the temperature of 1 gram of water 1 °C at atmospheric pressure. As a result, the specific heat of water is then 1 cal/g-°C. The specific heat is an indication of how much thermal energy a material can hold in its molecular structure for a given mass. As such, we will find that the specific heat of air is actually higher than the specific heat of concrete, on a gram for gram basis! Liquid metals tend to have the lowest specific heat of any substances, which is why mercury is used for thermometers – it requires the absorption of very little heat from its surroundings for its temperature to change.

Viscosity (μ) (units of lb/ft-hr, kg/m-hr) is defined as the ability of a fluid to resist flow. For example, if a force acted on

MATERIAL	SPECIFIC HEAT (J/g K)
Water	4.186
Wood	1.800
Air	1.0
Aluminum	0.9
Glass	0.84
Concrete	0.653
Steel	0.5

▲ **Figure 3-7** Specific heat of various materials

a high viscosity fluid such as molasses, the fluid would be much more resistant to moving than if the same force were applied to a lower viscosity fluid such as water. The viscosity of air is extremely low which explains why air in a room is a very poor insulator, while trapped air is one of the best insulators. Unconfined air is set into motion very easily, and thus quickly exchanges heat through convection, whereas trapped air cannot move, and thus can act as an insulator for reducing heat exchange by conduction. Note, however, that viscosity depends on the type of flow as well as the temperature and pressure, so fluids can quickly become more or less viscous depending on their state (this is another reason why the Navier–Stokes equations are so complex – many of the material properties are dependent upon the unknown variables of temperature and pressure).

Radiation

Radiation is the mode by which heat is transferred by electromagnetic waves, thereby *not* requiring a medium for transport; indeed thermal radiation can take place in a vacuum. Electromagnetic radiation, which is essentially the broadcasting of energy by subatomic transport processes, encompasses much more than just thermal radiation. All surfaces at a temperature above absolute zero (−460 °F or −273 °C) radiate thermal energy to other surfaces, but the amount they radiate is dependent on their temperature. Although low temperature surfaces will radiate to high temperature surfaces, the net difference in radiation will be from high temperature surface to low temperature in accordance with the Second Law of Thermodynamics. Radiation obeys the Stefan Boltzman law:

$$dQ/dt = \sigma(A_1 \times \varepsilon_1 \times T_1^4 - A_2 \times \varepsilon_1\varepsilon_2 \times T_2^4)$$

radiant exchange between two surfaces directly facing each other

σ is the Stefan Boltzman constant, and ε is the material property of emissivity.

Although the law is relatively straightforward physically, it is not practically solved. Radiation can travel enormous distances, and will continue to do so until it is interrupted by a surface. Any surface that is at an oblique angle to the radiation path will receive a reduced amount of radiation. As a result, view factors must be determined for every radiating object that a surface is exposed to, rapidly increasing the complexity beyond the idealized case, even though tempera-

MATERIAL	EMISSIVITY
Aluminum (anodized)	0.77
Aluminum (polished)	0.027
Steel (oxidized)	0.88
Steel (polished)	0.07
Glazed tile	0.94
Concrete	0.92
Glass	0.92
Brick	0.84
Paint, flat white	0.992
Paint, cadmium yellow	0.33

▲ **Figure 3-8** Emissivities of common building materials

ture is the only state variable. The design variables include the area and orientation of exposed surfaces, transparent as well as opaque. Emissivity is the primary material property affecting the rate of radiation transfer from the high temperature surface, and the property of absorptivity determines how much radiation the low temperature surface retains.

Emissivity (ε) (expressed as a unitless ratio from 0 to 1) is the measure of the ability of a surface to emit thermal radiation relative to that which would be emitted by an ideal 'black body' at the same temperature. The emissivity of a surface depends not only upon the material and temperature of the surface, but also upon the surface conditions. Scratched surfaces tend to have higher emissivities than polished surfaces of the same material at the same temperature.

Absorptivity (α) (expressed as a unitless ratio from 0 to 1) is the measure of how much thermal radiation is actually absorbed by a material relative to the total amount of thermal radiation that is incident on its surface. Related to absorptivity are **reflectivity** (ρ), which is the amount of thermal radiation reflected from the surface relative to total incident radiation, and **transmissivity** (τ), which is the amount transmitted through the material relative to the total. All three ratios are related as follows: $\alpha + \rho + \tau = 1$.

These three modes of heat transfer determine the location, direction and timing of all movement of heat, whether from animate or inanimate objects, within any thermal environment. And, all of these modes take place at the boundary between a material system and its surroundings. Within most air environments in buildings, all three modes of heat transfer will contribute to producing the heat exchange between an entity and its local surroundings, rendering the quantitative determination of the air conditions beyond the scope of most sophisticated numerical simulation codes.[4] For example, a heated wall will radiate to other cooler walls in view; it will also transfer heat through conduction to anything in direct contact with it, including the immediately adjacent air which will then start moving as it heats up, leading to further energy exchange to the remaining room air through convection.

Scale further differentiates these behaviors from each other. While these behaviors can occur over large distances and in large volumes, each has a characteristic scale at which their boundaries can be manipulated. Conduction, as the transfer of energy through the direct exchange of kinetic energy from molecule to molecule, can be best controlled at the meso-scale. Radiation, including light, is dependent upon the physical characteristics of surfaces, for example, polished aluminum has a lower emissivity than etched aluminum, and thus is a micro-scale behavior. Convection, which also explains sound transmission, requires the movement of a fluid, driving the scale to centimeter-size and above.

Thermodynamic scale	Length scale (meters)	Boundary process
Macro-scale	cm to m+	Convection
Meso-scale	mm to cm	Conduction
Micro-scale	μm to 0.1 mm	Radiation
Nano-scale	pm to nm	Non-continuum

THE THERMAL ENVIRONMENT OF THE BODY

Our ultimate goal as designers is to provide for the health, welfare and pleasure of the human body. The human body does more than its share in maintaining its own health. An intricate and versatile thermoregulatory system can accommodate an astonishing range of environmental conditions – the peripheral skin temperature alone can vary from about 10 to 42 °C without harmful consequences. The term homeostasis – the maintenance of a stable body temperature – is a bit of a misnomer, as it is only the temperature of the internal organs that must be maintained at a consistent level. The rest of the body functions as a heat exchanger, dynamically utilizing radiation, conduction, convection and evaporation to adjust the body's thermal balance. A body in thermal equilibrium with its environment, defined as no difference between stable body conditions and stable surroundings, is not animate. Nevertheless, the objective for HVAC system design has been to establish a neutral environment – one in which '80% are not dissatisfied':[5] phenomena that we can't see, to produce sensations that we can't feel.

Thermal sensation is yet another differentiating aspect of the human nervous system, and, furthermore, it is not directly linked to the body's thermoregulation as is commonly assumed. The cutaneous receptors (or what we traditionally call 'touch') respond to two large classes of environmental stimuli – mechanical and electromagnetic energy. These receptors – known as mechanoreceptors and thermoreceptors – are excellent examples of boundary crossing in our thermodynamic system because they respond only to stimuli at the interface between our body and its surroundings. We recall, however, that there must be a difference in one of the state variables for energy to cross a boundary. As such, thermoreceptors do not sense ambient temperature at all, but

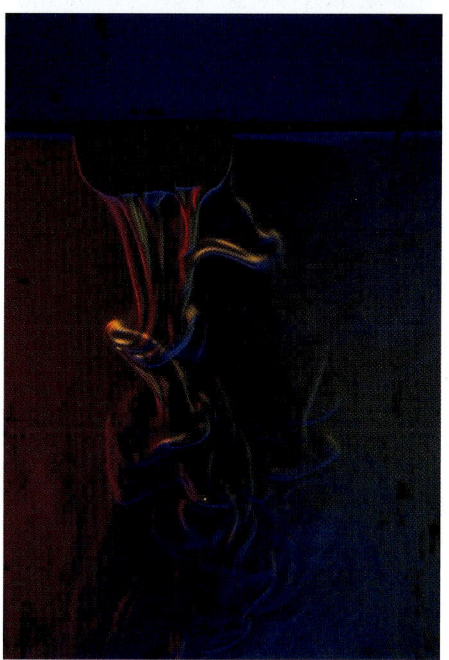

▲ **Figure 3-9** Images of buoyant convection. The top image shows the buoyant plume above a candle flame, and the bottom image shows the downward plume as an ice cube melts

rather they respond to the difference between our skin temperature and its surroundings. Skin temperature is one of the most variable of all of the body's thermal regulation responses, and so we can assume that the difference is continuously shifting. Our lack of awareness of this constant adjustment of our thermal state is not due to the homogeneity of the surroundings; rather it is an indication that change is the normative state in the neurological system. The thermoreceptors do not produce a consciously aware sensation until the derivative of the change – the rate – begins to change. We might say that we only become aware of our surroundings when there is a 'difference' in the difference. The body is not a thermometer.

The human body is the most typical of the heat exchanging entities within a building. If we characterize the building environment by the thermal phenomena commonly taking place, and not by the HVAC technology used to mitigate those phenomena, we will recognize that all of the phenomena result in buoyant behavior. Buoyancy occurs when gravity interacts with density. For example, we know that warm air rises and cool air sinks. Air density is inversely proportional to temperature, so as the temperature rises, the density drops. The action of gravity pulls the denser air toward the ground resulting in a vertical stratification of temperature from low to high as the elevation above the ground plane increases. The buoyant plume that surrounds the body is also found around other heat sources in the building – lighting, computers, electrical equipment – as well as around many processes – cooking, heating, bathing. Any entity that produces heat within surroundings of air will exchange its heat through buoyancy. In addition, any time there is a difference in temperature between a surface entity and the surrounding air, there will be a buoyant boundary layer. The surface temperatures in a building, particularly those on exterior-facing components such as walls, windows, roofs and floors, are almost always different from the ambient air temperature, thus producing buoyant flow along surfaces. The interior thermal environment, rather than being a singular bounded state, is a large collection of buoyant behaviors, all of which have unique boundaries.

The HVAC system of today, and of the previous century, mixes and then dilutes these multiple energy systems for the purpose of controlling the temperature of the entire air volume. This is undoubtedly one of the least efficient ways of managing the human thermal balance. Compare this approach to another type of response to a common buoyant boundary layer problem – that of aerodynamic lift. Subtle and often microscopic modifications in the surface of an airfoil can dramatically affect the boundary layer conditions between the airplane wing and the atmosphere. If one treated this energy exchange problem in the same manner as we use for mitigating the energy exchanges in a building, then we would be trying to manage the pressure of the entire atmosphere rather than that within a few centimeters of the plane's surface. Ridiculous, yes, but this is yet another example of the peculiar relationship between science and building technology that we discussed in Chapter 1. In aerodynamics, the technology is developed and modified to respond to particular problems of physics. In building design, we modify the environment (the physical behavior) to optimize the performance of the technology.

Action at the most strategic, and efficient level, requires knowledge of where the energy transactions naturally occur and an understanding of their scale. HVAC systems are designed in relation to the scale of the building, whereas thermal behaviors operate at much smaller scales. The ideal response will occur at the boundary and scale of the behavior. Smart materials and new technologies – due to their small scale – will eventually provide the direct and local action that will allow us to design a thermal environment rather than only nullify our surroundings.

3.6 The luminous environment

Both light and sound are thermal energies – light as electromagnetic radiation, and sound as pressure-driven convection. As such, we can place the discussions of luminous and acoustic behavior within the same context as those surrounding the thermal environment. Light, however, has meaning only for animal perception, and for our purposes, this narrows down to human perception. Indeed, the definition of light is 'visually evaluated radiant energy'. Radiant energy, or electromagnetic radiation, is energy movement through space in the form of oscillating or fluctuating electric and magnetic disturbances. The radiation is generated when an electrical charge accelerates, such as what occurs as a result of the rapid oscillation of electrons in atoms. The oscillations release periodic 'packets' of energy or 'photons' that travel away from the vibrating source at the speed of light in a sinusoidal pattern. Although the term photon is derived from the Greek word *photos* for light, the entire electromagnetic spectrum is comprised of photons, from gamma rays to microwaves.

Electromagnetic radiation can be characterized by its energy (E), wavelength (λ – distance from wavecrest to wavecrest) and frequency (ν), all of which are interrelated in the following two equations.

$\lambda = c/\nu$
where c = the speed of light (299,792,458 m/s)

$E = h \times \nu$
where h = Planck's constant (6.626×10^{-27} erg-seconds)

The electromagnetic spectrum encompasses wavelengths as large as the height of a mountain and as small as the diameter of an atomic particle – a span of about 15 orders of magnitude. Within this enormous range of energies, light occupies an almost negligible band of wavelengths – from about 400 to 750 nm – or less than 0.0000000000000000003% of the spectrum! Light is the physical phenomenon most responsible for our perception of the world, and yet it is an almost negligible fraction of the electromagnetic energy that surrounds us and connects us to all other things in the universe.

A good rule of thumb regarding electromagnetic radiation is that its interface boundary is on the same scale as its wavelength. For example, radio signals have wavelengths from about 100 to 1000 meters, with FM having a shorter wavelength than AM. As a result, mountain ranges are much more likely to interfere or interact with FM, whereas buildings have less impact on its signal. Infrared radiation, with a wavelength in the micron to millimeter size, is most affected by an object's surfaces. At the other end of the spectrum, gamma rays with wavelengths on the order of 10^{-12} meters can penetrate surfaces and molecules, acting on the atomic scale. The narrow band of light is most effective at the scale of surface *features*. The surface features of an object tell us many things about it – its color, its texture, its orientation, its shape – basically providing us with the necessary information for negotiating through the world of physical objects.

This enormous range of wavelengths, from as large as kilometers to as small as the diameter of an atomic particle, has led to three different models for describing the behavior of electromagnetic radiation. The first two models span the entire spectrum, and together they produce 'wave-particle duality'. Photons are differentiated by amount of energy only, yet 'world-view descriptions' of their behavior will attach either wave-like or particle characteristics to them, the former to low energy photons and the latter to high energy photons. Although photons are both – a discrete packet traveling in

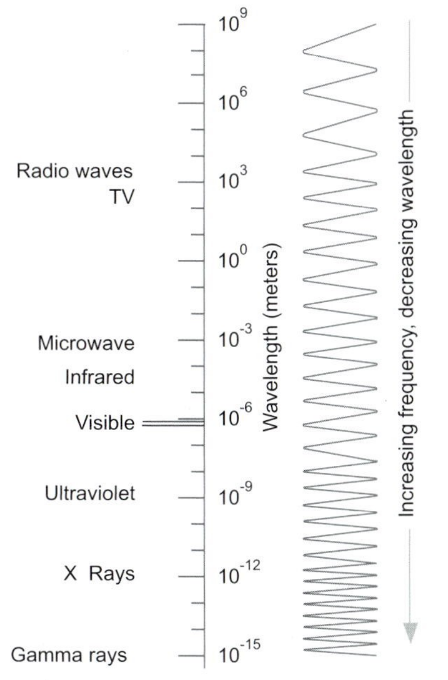

▲ **Figure 3-10** The electromagnetic spectrum. Light occupies a tiny portion from 0.4 to 0.75 microns (10^{-6} meters).

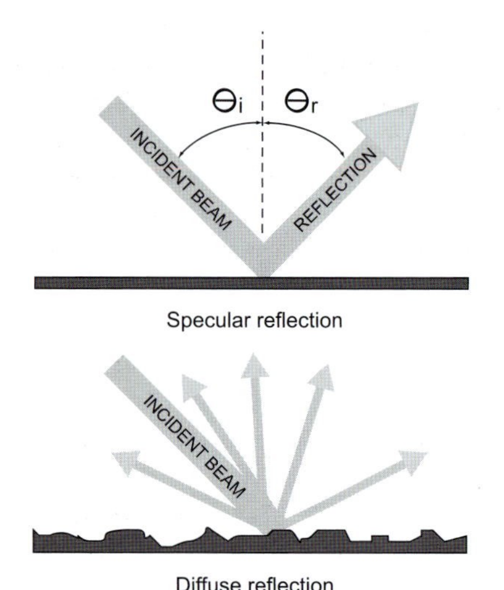

▲ **Figure 3-11** The law of reflection states that the angle of reflection must equal the angle of incidence. This is true even for a diffuse surface, as the angles are determined from the surface features, not the plane

MATERIAL	REFLECTIVITY
Aluminum (etched)	0.8
Aluminum (polished)	0.65
Aluminum (brushed)	0.55
White plaster	0.91
White terracotta	0.7
Stainless steel	0.55
Chrome	0.6
Light wood	0.6
Limestone	0.45
Concrete	0.2

▲ **Figure 3-12** Reflectivities of common building materials

wave motion – these descriptions help to simplify the relevant phenomena at the scales in question. The third model is only used to describe the behavior of light. Geometric optics has none of the characteristics of waves or particles, yet it is a very effective model for determining the path of light. We can isolate four rules of geometric optics to help us predict where light will travel.

1. Light travels in a straight line between two points (surfaces).
2. When light strikes a surface, it can be absorbed, transmitted and/or reflected. For any given surface, the amounts of each are determined by the ratioed material properties of reflectance (*r*), transmittance (*t*) and absorptance (*a*), such that:

   $$r + t + a = 1$$

   The three properties depend on the surface qualities as well as the molecular structure of material. Because the wavelength of light is so small, extrinsic changes in the surface structure can dramatically affect the material's interaction with light. For example, the mechanical working of a surface through cold rolling will tie up the free electrons on the surface of a metal, thus reducing its ability to re-emit electromagnetic radiation. As a result, a material like aluminum can have a reflectance of 0.8 if it is etched, but 0.6 if the surface is polished. Furthermore, while these three properties describe the disposition of all electromagnetic radiation incident on a surface, their specific values are wavelength-dependent.
3. If light is reflected, it will reflect from a surface at the same angle as it arrived but in the other direction. This is also known as the **law of reflection**:
   The angle of incidence of radiation is equal to the angle of reflection: $\theta_I = \theta_R$

   Diffusion versus specularity: Specular surfaces are microscopically smooth and flat such that the plane of any surface feature lies in the same plane as the overall surface. Mirrors and highly polished surfaces tend to be specular. Diffuse surfaces have surface irregularities that do not lie in the same plane as the overall surface – the law of reflection still applies, but the angle of incidence is particular to the surface feature that each photon is incident upon, thus resulting in the scatter of light.
4. If light is transmitted, it will refract at an angle related to the ratio of the refractive indices of the two media. This is also known as the law of refraction: When light passes

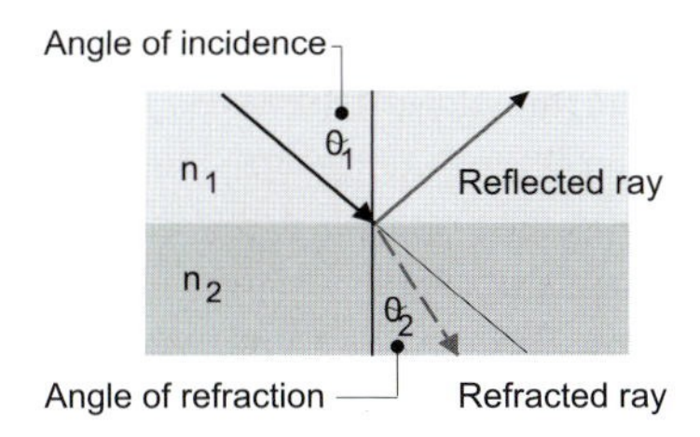

▲ **Figure 3-13** Snell's Law. Snell's Law determines the angle of refraction when light passes from one transparent medium to another

from one medium to another, its path is deflected. The degree of deflection is dependent upon a material property known as *the index of refraction (n)*. The value of n is measured with respect to the passage of light through a vacuum, and, as a result, all transparent media, from air to diamond, have indices of refraction greater than one. The amount of deflection is determined from the following relationship:

$$\sin\theta_1 = n_{21} \sin\theta_2$$

where n_{21} = the index of refraction of medium 2 with respect to medium 1.

Only when the incident light is normal to (perpendicular to) the surface will the path angle continue in a straight line.

Critical angle: If the refractive index of the material that light is transmitting from is larger than the refractive index of the refractive index of the material that light will be transmitting into, then there exists a critical angle beyond which light will not transmit but is reflected internally back into the first material. The critical angle is defined by the following relationship:

$$\sin\theta_c = n_1/n_2$$

where n_1 = refractive index of outside material and n_2 = refractive index of starting material.

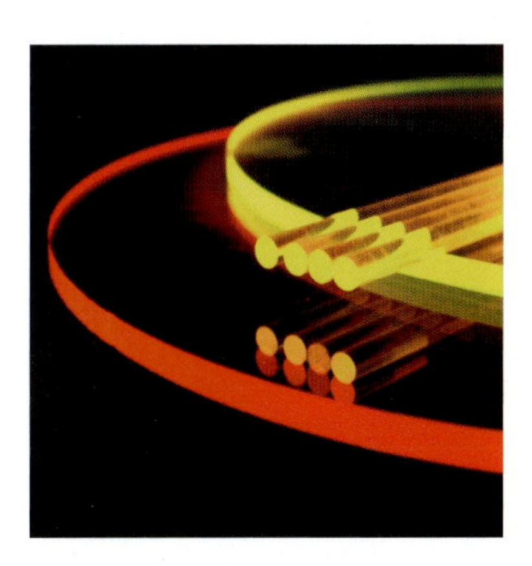

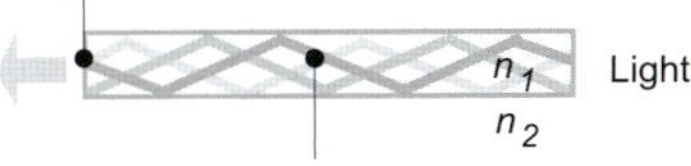

▲ **Figure 3-14** Internal reflection. Light is bounced back and forth internally in the material and emerges at the edges

QUALITIES OF LIGHT

Many of the material properties that interact with light are extrinsic, in that conditions other than the molecular makeup of the material will affect the property. Reflectance, absorptance and transmittance fall into this category. We have already talked about how alterations in the surface structure can have a large impact on these properties, but even more influential are two key parameters of light – its intensity, or energy, and its spectral composition. The intensity is the amount of photons per unit area in a particular direction. A common analogy for describing intensity involves a garden hose. When the nozzle of the hose is rotated in one direction, a narrow and high pressure stream emerges. When the nozzle is rotated in the other direction, a fine mist will fan out. The former is high intensity, the latter is low intensity. The intensity will ultimately determine how our visual system perceives an object in relation to its surroundings.

The sun is the primary source of light for us, and it also sets the standard of comparison for artificial light sources. While all objects produce radiation, they do so in different parts of the spectrum. The sun emits radiation energy primarily (95%) in a range of wavelengths from about 2×10^{-7} to 4×10^{-6} m, peaking in the visible part (45% of the solar radiation) of the electromagnetic spectrum, and extending into the ultraviolet (10%) and near infrared (45%) regions. The relative continuity of its intensity levels throughout the visible part of the spectrum provides an even level of illumination on surfaces, allowing them to reflect any colors determined by their surface features. Conversely, the early discharge lamps tended not only to be discontinuous across the spectrum, but also had energy levels concentrated in narrow bandwidths.

Even though we routinely talk about color as if it belongs to objects – blue water, green grass, red wagon – color belongs only to light. All surfaces are subtractive in that they can only subtract energy and color from light, not add to it. For example, the spectral distribution of a low-pressure sodium lamp has a very narrow bandwidth, with wavelengths confined to the yellow range. When a blue car parks under this lamp, the surface of the car can only reflect what is provided to it, and in this case it can only reflect yellow. At best, the car will appear to be a dark brown, and it may even appear to be black if there are no yellow components in its paint. If the same car is still outside the next afternoon, it will appear as the blue that was intended when the paint color was selected.

We can describe the color of sources and the reflected color of objects with three quantities – energy, wavelength and bandwidth. The energy level tells us how bright, the wavelength tells us which hue and the bandwidth tells us with what saturation. Compare the following two spectral profiles. The first one is that of a laser. The energy level is high, indicating that the light will be extremely bright; the wavelength is centered on 640 nm, producing red, and the bandwidth is very narrow, suggesting to us that the red is a very pure red with no other wavelengths involved. The second profile belongs to an incandescent lamp. Although it, too, peaks at a red wavelength, the distribution of its profile across the entire visible spectrum indicates that it is unsaturated, containing many more wavelengths than those in the red part of the spectrum. As such, an incandescent lamp does not appear to be red to us, although we do recognize its white light as being 'warmer' than that produced by daylight. In addition, its low energy levels indicate that it is not delivering very much light, which concurs with our expectation that a laser produces much brighter light than does our bedside lamp.

The color of a surface can be described in the same way as for the source by a spectral profile. For example, we could also compare the spectral profiles for a tomato against those of a blue pigment. The key difference between surface color and source color is that there must be a fairly specific match between the surface and source profiles for the color to appear as intended (either by nature or by the designer!). A source with long wavelength light will render the tomato fairly accurately, but the blue pigment may appear to be totally absorptive (black) if there are no short wavelength components in the light source. For this reason, we often choose the neutrality of a continuous spectrum white light to ensure that colors are rendered accurately. Light that reflects off a surface then becomes a source as well, but a diminished one. Besides the reduction in energy, and the subtraction of specific wavelengths, a surface can impact one more quality of light, that of polarization.

Returning to our earlier discussion where we introduced the concept of the electromagnetic wave, we need to be aware that these energy pulses oscillate in three dimensions and not in two dimensions, as they are usually depicted. Each atom in a source will emit light in a different plane. As a result, the sun and many other common light sources produce photons that oscillate in planes randomly oriented to one another – a condition called 'unpolarized light'. When there is a preferential orientation to the planes, the light is said to be 'polarized'. While this distinction is of crucial importance in many optical phenomena, it is interesting to note that the human eye cannot normally distinguish between polarized and unpolarized light.

The condition of polarization can occur for many reasons. Many materials produce preferred directions for electric fields. When light passes through them, the internal structure of the material naturally produces polarized light. Absorption or reflection can be higher for one direction of polarization than another. Calcite, for example, is a naturally found material that produces polarized light as it passes through. More generally, any material that exhibits the property of 'dichroism', the ability of the material to absorb light vibrating in one orientation more strongly than in the other direction, can be used as a polarizing material. Tourmaline, for example, is a natural dichroic crystal that has traditionally been used as a polarizer. Many synthetically produced materials also produce this same effect. They often contain long rod or plate structures that are in a regular arrangement. These aligned

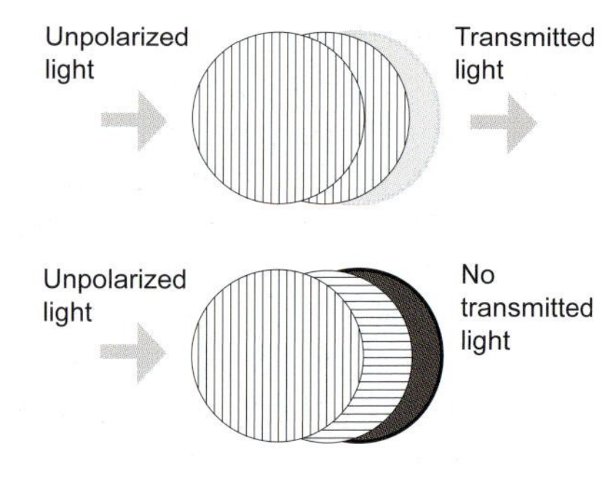

Unpolarized light becomes polarized as it passes through the first plate. Depending on the orientation of the second plate, the polarized light may pass through or be blocked.

Parallel polarizing plates

Crossed polarizing plates

Colored fringe patterns show up in birefringement materials placed between plates

▲ **Figure 3-15** Polarized light

structures can absorb one plane of polarized light while transmitting the other plane.

The selective properties that different materials have for producing polarized light can be used in many ways. When a polarizing material that transmits light that is only vertically polarized is exposed to light that is horizontally polarized, no light will be transmitted through the material at all (all horizontally polarized photons trying to pass through will be stopped). This situation is commonly exploited in 'polarized' sunglasses. When sunlight is reflected from a horizontal surface, including water and snow, it becomes partially horizontally polarized. Sunglasses with vertically oriented polarizing materials can block this reflected light, thus reducing glare.

THE LUMINOUS ENVIRONMENT OF THE BODY

This is not enough background, however, to explain the luminous environment. Just as our skin operates as the boundary between our body and the thermal environment, then so do our eyes with respect to the luminous environment. More specifically, that boundary is located near the back of our eye within the tiny region composed of our visual receptors – the rods and cones. Like any other surface, these receptors will selectively absorb certain wavelengths at certain energy levels. As children, many of us were taught about rods and cones, the rods serving for night vision and the cones for color. Wavelengths were attached to these, and we assumed that the cones were red, green and blue and that the rods saw only black and white. Advances in neurology and physical psychology during the past decade have given us a very different 'view' of the photoreceptors in the eye. The peak wavelengths for all of our receptors reside in the shorter to middle range of the visible spectrum – the three cones peak at 420, 530 and 560 nm and the rods peak at 500 nm. Essentially, our visible system is most sensitive to green.

Current models of the eye separate its neurological response into two major categories: the 'what' system and the 'where' system which together replace the older rod/cone system.[6] These two categories are associated with two different types of ganglion cells, with the larger cells producing the 'where' response and the smaller cells producing the 'what' response. The fundamental purpose of both types of ganglion cells is to establish relative comparisons of photon reception between small areas of the retina. Most of the comparisons take place through a center-surround receptor field – in the center of the field photons excite the cell and in the surround of the field, photons inhibit the field. As a result, a constant light level across the field produces a null signal, regardless of how light or dark the level may be. Just as the body is not a thermometer, the eye is not a light meter. Only when the receptor field encounters a difference in the photons across the area does it signal the brain.

In the 'where' system, these differences are responsible for the perception of motion, depth and spatial organization, as well as the segregation of figure/ground. The 'where' system is color blind, but is highly sensitive to differences in luminance, or contrast. Conversely, the 'what' system is highly color selective, but is relatively insensitive to luminance contrast. This system is responsible for object and face recognition, and, of course, for color perception. Acuity is highest in the 'what' system, but the 'where' system is faster, making it ideal for perceiving motion.

This new understanding of the visual system has profound implications for designers and particularly for architects. If

▲ **Figure 3-16** Gray bar sequence. The center bars in both images have identical luminances, only the background is different. (Image courtesy of John An)

luminance alone is responsible for the determination of where something is, then we have the possibility of creating visual articulation of a surface where there is none, as well as vice versa. If color alone is responsible for object recognition, then similar objects can be further differentiated by a planned use of color. We will have the unprecedented ability to design how someone sees and interprets information, as opposed to designing only what is placed in front of them.

3.7 The acoustic environment

The thermal environment in a building may have been coerced into neutrality, and the luminous environment is generally an afterthought, but the acoustic environment has been well documented and explored since antiquity. Nevertheless, the understanding of acoustics did not arrive before our understanding of heat and light, indeed it was quite late, not until the turn of the 20th century. The current model for how the ear works was not developed until the end of the 20th century, and it is well accepted that the neurology of the ear is still not as well understood as that of the eye. Why, then, is this the one environment that architects have developed tremendous expertise in designing?

The answer may well be in the scale of the thermal behavior that determines the transmission of sound. Sound, which is produced by pressure pulses in a fluid medium, is transmitted by convection. As we discussed before, convection operates at the largest scale of any of the thermal phenomena. Because it is the same scale of architectural objects, there has always been a direct and immediate connection between architectural objects and the reception of sound. If we cover a concrete wall with wood paneling, we will change the acoustic qualities in the space in a very predictable way. We have developed ways of predicting similar types of behavior in thermal and luminous environments, but the thermal predictions depend heavily on empirical observations, and the light predictions cannot take into account the microscopic behavior of light as it interacts with surfaces and our eye. Furthermore, a long-standing type in architecture is the theater, a space whose design has been more dedicated to acoustics than to any other aspect. There is no such corollary for the other two environments.

Like light, sound is thermal energy that also can be characterized by wave-like behavior. Sound is produced by mechanical (kinetic) energy that is propagated through an 'elastic' medium by vibration of the molecules of the medium. By elastic, we are referring to any medium that has a

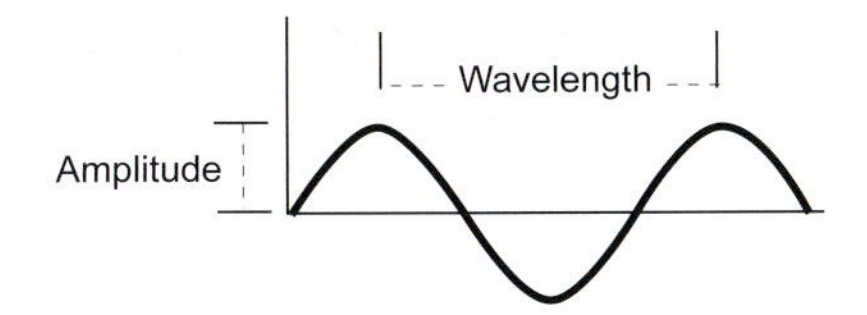

▲ **Figure 3-17** Sound wave. Sequential compressions and rarefactions will produce an oscillating pressure field. The amplitude is an indication of how loud a sound is, and the wavelength is an indication of its frequency, or pitch

compressible component; fluids such as air are obviously elastic, but solid substances such as concrete also contain interstitial air spaces that can propagate sound. The origin of sound can be any disturbance (also known as a source) that produces a displacement of the surrounding medium. This may be the mechanical impact on a solid body, oscillating air pressure released by a whistle or horn, or electrical energy acting on a membrane causing it to deflect. The resulting disturbance will cause successive compressions and rarefactions in the medium that will radiate spherically in the form of waves from its origin. If light can be described as a series of electromagnetic energy pulses, then sound could be described as a series of pressure pulses. Sound waves are characterized by their frequency, wavelength, pressure (amplitude) and phase. Wavelength and frequency are inversely related in the following equation:

$$V = f \times \lambda$$

velocity of sound in the medium (m/s) = frequency (cycle/s) × wavelength (m/cycle)

This equation should be recognizable as the same one shown for light with the speed of light replacing the velocity of sound. The primary difference is that while the speed of light is a constant, the speed of sound is dependent upon the medium: both its composition and its state. For example, the speed of sound in air is about 345 m/s (depending upon air temperature and pressure), but it is more than four times as fast when the medium is water (1450 m/s) and almost twenty times faster in granite (6000 m/s). If we compare these velocities to that of light, 300 million m/s, we recognize that sound is extremely slow. We are unable to distinguish when light was generated, as light from the sun reaches us infinitesimally close to when light from our table lamp reaches us. But sound's slowness is omnipresent. We know that counting the seconds between seeing a lightning strike and hearing the rumble of its thunder will tell us how far away it is. We have noticed the delay between the motions of a marching band and when we finally hear their music. The distance between the source and the intended receiver looms as an important design variable. The distances involved can be as small as room scale, which is certainly the main reason that architecture and the acoustic environment have been so intertwined for the past two millennia. Room proportions directly determine the loudness of sound, and materials determine its clarity. As a result, while we can apply many of the observations about light to sound, we must include several others that will directly influence architectural design.

As a radiant transport phenomenon, sound shares many important physical characteristics with electromagnetic radiation:

- Both are transmitted by waves or wavelike motions, and their path of transmission obeys geometric optics.
- Both radiate spherically from their source, with their intensity falling off with the square of the distance from the source.
- The processes of transmission, reflection and absorption apply to both sound and light.
- Both sound and light travel at a speed that is nearly independent of frequency and wavelength (we see the red components of the spectrum at the same time we see the blue, we hear high frequency sounds at the same time we hear low frequency sounds).

These similarities led to the use of sight lines for determining sound propagation from a source, and, today, many of the acoustic simulation tools use the same ray tracing techniques that were developed for light simulation. A common rule of thumb is that if you can see the source, then you can hear it. For centuries, theater designs were based on this accepted rule, from the tiered steps of the classical amphitheater to the horseshoe shape of the Baroque opera house. The development of the science of acoustics at the turn of the 20th century was predicated on characterizing the sound behaviors that were *not* like those of light.

Replacing geometric form as the determinant of acoustic design, materials emerged as the predominant factor. This influence comes from the multiple roles played by the material property of absorptivity, which is an indication of how much kinetic energy the material can absorb from the pressure pulses, thereby diminishing their amplitude. The kinetic energy arriving at a surface can be quantified as the sound intensity, which is the magnitude of acoustic energy contained in the sound wave. The sound intensity is proportional to the amplitude of the pressure difference above and below the undisturbed atmospheric pressure. Because we are most interested in human environments, we will use a modified version of sound intensity known as the sound intensity level. Whereas sound intensity is an objective measure of energy, the sound intensity level is a subjective measure which takes into account the sensitivity of the human ear. Fechner's Law states that the intensity of sensation is proportional to the logarithm of the stimulus. Sound intensity level, in decibels or dB,[7], then relates logarithmically to the human hearing experience and is expressed by the following equation:

$$\text{IL (intensity level in dB)} = 10 \log (I/I_0)$$

where I = sound intensity, and I_o = the sound intensity of the quietest sound the human ear can hear.

The material property of absorptivity affects the intensity level in three ways. The inverse square law causes sound levels to drop off quite suddenly with distance. In large spaces, this would be highly problematic, as listeners in the room may or may not be able to hear the source. Reflections are needed to amplify the sound so that any single listener will receive direct sound plus any closely following reflections of that sound. The absorptivity of the materials in the space determines which frequencies are reflected and in which direction, and can also be utilized for canceling unwanted and slow reflections that might cause an echo.

In acoustic design, we often use a modified version of the absorptivity. The sound absorption coefficient is the ratio of the sound-absorbing effectiveness of a unit surface area (1 m^2) of a given material to the same unit surface area (1 m^2) of a perfectly absorptive material. Represented by the Greek letter α, the sound absorption coefficient is usually expressed as a value between 0 and 1, where 0 is for a perfectly reflective material, and 1 is for a perfectly absorbing material. Note that a perfectly absorbing material may also be highly transmissive, for example, an open window has a coefficient of 1.

α = sound energy not reflected from material/sound energy incident on material

(Because we most typically deal with air, we will often find that materials that are good thermal insulators are also good acoustic absorbers. Sound absorbing materials fall into two generic types: porous absorbers and resonant absorbers. Porous absorbers have interstitial spaces where viscous flow restrictions through the pores reduce the sound energy. Resonant absorbers act as a mass and a spring by absorbing energy and resonating back at a particular frequency.)

Interior spaces, unless engineered to be fully absorptive, will often have many diffuse reflections. These will give more body to the sound, but will also raise the ambient sound level in a room. The absorptivity of the materials will ultimately determine the ambient or background sound level. We can determine the impact of changing or adding materials in any given situation with the following equation:

$$Il_2 = Il_1 - 10 \log A_2/A_1$$

where Il_1 is the starting sound level, Il_2 is the final level, $A = \Sigma\alpha_i \times$ surface area$_i$.

A room with many hard, hence reflective, surfaces such as concrete and stone will have a much higher background sound level than a room filled with good absorbing materials such as upholstery. Ultimately, if completely efficient sound-absorbing materials are placed on all boundary surfaces of a room, outdoor conditions will be approximated where only the direct sound remains.

Most significant, however, for the development of the modern science of acoustics was the discovery by Wallace Sabine in the late 19th century that material absorptivity impacted the reverberation time of a room. Reverberation is the continuation of audible sound after the sound source is cut off. If we had materials that were perfect reflectors, the sound would never die down. The amount of time that a sound persists is known as the reverberation time. The definition of reverberation time is the amount of time that elapses before there is complete silence after a 60 dB sound has stopped (or the amount of time it takes for a sound to decay by 60 dB, to a millionth of its original sound intensity). A space is considered to be live if it has a long reverberation time, and dead if it is short. Organ music was developed for the long reverberation times of cathedrals, whereas speech needs a room that has a very short reverberation time. We can calculate reverberation time for a space using Sabine's formula:

$$T_r \text{ (in seconds)} = 0.16 \text{ Volume}/A \ (\Sigma\alpha_l \times \text{surface area}_i)$$

If the understanding of the impact of absorptivity has given the designer great control of the acoustic environment in a space, recent developments in electro-acoustics now allow the ability to design acoustic environments independently of the physical surfaces of architecture. Sophisticated signal processing and the selective placement of micro-speakers can remove much of the macro-scale influence of architecture on sound by accomplishing tasks such as reflective amplification, reverberation, diffusion and sound direction electronically. Performance halls that must accommodate a multitude of acoustic requirements, from those of a lecturer to that of a full symphony orchestra, were among the first adopters of electro-acoustics. Touring shows often bring and install their own electronic systems to ensure quality control of the acoustics regardless of the venue. Although many acousticians claim that they can hear the difference between a 'live' hall and an electronic hall, the rapid evolution in the micro-technology coupled with advances in sound simulation will soon bring comparable performance to electro-acoustics.

THE ACOUSTIC ENVIRONMENT OF THE EAR

Perhaps more has been known about the acoustic environment than any of the other two environments, but much less is known about how the ear responds than how the eye and the skin respond to stimuli. Only in the past 20 years have the roles of the two primary mechanoreceptors in the ear been identified, and their specific functionality is still being verified. Unlike the eye, in which there is a one-to-one mapping of photons to photoreceptors, the mechanoreceptors must respond simultaneously to overlapping frequencies, amplitudes and directions of sound. Furthermore, whereas the eye has approximately 150 million receptors, the ear must perform its more complex role with only 20 000 receptors. Although there is universal agreement that the hair cells are the key to understanding the sensitivity of the ear, there is as yet no coherent theory on just how they work.

The characteristic that we are most interested in as designers is how the ear spatializes sound. A large amount of our awareness of the space surrounding us comes from non-visual stimuli. Proprioceptors in our lower body give us a sense of how close or far from a wall we might be, while the mechanoreceptors in the ear give us the cue as to how spacious a room is. Just as the Ganzfeld effect, by eliminating luminance contrast, erases any visual comprehension of the dimensions of a space, so too does an anechoic chamber in regard to sound. Without a sonic feedback from our surroundings we are incapable of placing ourselves spatially in a room even if its walls are clearly defined visually. Many installation artists are beginning to experiment with sonic manipulation, creating spaces where there were none, and directing the localization of sound at will.

Smart materials, in the form of piezoelectrics, are already playing the central role in sound design, but the potential of *designing* the acoustic environment, as well as the thermal and luminous environments, directly may well be the most provocative application of smart materials in the design field.

Notes and references

1. For an overview of the historical development of the Law of Thermodynamics, see D.S. Cardwell, *From Watt to Clausius: The Rise of Thermodynamics in the Early Industrial Age* (London: Heinemann, 1971). Chapter 8, 'The New Science', discusses Clausius and the understanding of entropy.
2. Fitch, J.M. (1972) *American Building 2: The Environmental Forces That Shape It*. New York: Schocken Books, page 1.
3. For an in-depth review of the 19th-century determinants of today's HVAC system, see D. Michelle Addington, 'Your breath is your worst enemy', in *Living with the Genie*, eds A. Lightman, C. Dresser and D. Sarewitz (Washington, DC: Island Press, 2003), pp. 85–104.
4. Numerical simulation of building environments is still in its infancy. For a discussion of some of the issues with simulation, see D.M. Addington, 'New perspectives on CFD simulation', in *Advanced Building Simulation,* eds A. Malkawi and G. Augenbroe (London: Spon Press, 2004).
5. The definition of comfort is an elusive target, ranging from the broadly, and vaguely, qualitative definition from ASHRAE: 'Thermal comfort is the condition of mind that expresses satisfaction with the thermal environment', to the complex empirical derivation from Ole Fanger that includes calculations about the influence of clothing.
6. This terminology was developed by Margaret Livingstone, a Harvard University neurobiologist, and is fully explained in her landmark book, *Vision and Art: The Biology of Seeing* (New York: Harry N. Abrams, Inc., 2002). All of the following discussion regarding the 'what'/'where' system can be sourced to this book.
7. Representations of the sound intensity level are often in units of dBA rather than of dB. This is a weighted measure. The human ear discriminates against low frequency sounds such that a given sound level will appear to be louder in the mid- to high-frequency range than at the same level at lower frequencies. Weighting approximates this characteristic. dBA scales are weighted to correspond to the way the human ear works; dBC is relatively unweighted, dBB is an intermediate scale and dBD is a specialized weighting for aircraft noise.

4 Types and characteristics of smart materials

4.1 Fundamental characteristics

This chapter first identifies characteristics that distinguish smart materials from other materials, and then systematically reviews many of the more widely used ones. We begin by noting that the five fundamental characteristics that were defined as distinguishing a smart material from the more traditional materials used in architecture were transiency, selectivity, immediacy, self-actuation and directness. If we apply these characteristics to the organization of these materials then we can group them into:

1. Property change capability
2. Energy exchange capability
3. Discrete size/location
4. Reversibility

These features can potentially be exploited to either optimize a material property to better match transient input conditions or to optimize certain behaviors to maintain steady state conditions in the environment.

As we begin to explore these distinguishing characteristics, we will see that the reasons why smart materials exhibit these and other traits is not easy to explain without recourse to thinking about both the material science precepts noted in the last chapter and the specific conditions surrounding the placement and use of the material. Of particular importance is the concept of the surrounding energy or stimulus field that was discussed in Chapter 3. We recall that energy fields can be constructed of many types of energy – potential, electrical, thermal, mechanical, chemical, nuclear, kinetic – all of which can be exchanged or converted according to the First Law of Thermodynamics (the law of the conservation of energy).

The physical characteristics of smart materials are determined by these energy fields and the mechanism through which this energy input to a material is converted. If the mechanism affects the internal energy of the material by altering either the material's molecular structure or microstructure then the input results in a *property change* of the material. If the mechanism changes the energy state of the material composition, but does not alter the material, then the input results in an *exchange of energy* from one form to another.

A simple way of differentiating between the two mechanisms is that for property change type, the material absorbs the input energy and undergoes a change, whereas for the energy exchange type, the material stays the same but the energy undergoes a change. We consider both of these mechanisms to operate at the micro-scale, as none will affect anything larger than the molecule, and furthermore, many of the energy-exchanges take place at the atomic level. As such, we cannot 'see' this physical behavior at the scale at which it occurs.

Property change

The class of smart materials with the greatest number of potential applications to the field of architecture is the property-changing class. These materials undergo a change in a property or properties – chemical, thermal, mechanical, magnetic, optical or electrical – in response to a change in the conditions of the environment of the material. The conditions of the environment may be ambient or may be produced via a direct energy input. Included in this class are all color-changing materials, such as thermochromics, electrochromics, photochromics, etc., in which the intrinsic surface or molecular spectral absorptivity of visible electromagnetic radiation is modified through an environmental change (incident solar radiation, surface temperature) or a direct energy input to the material (current, voltage).

Energy exchange

The next class of materials that is expected to have large penetration into the field of architecture is the energy-exchanging class. These materials, which can also be called 'First Law' materials, change an input energy into another form to produce an output energy in accordance with the First Law of Thermodynamics. Although the energy conversion efficiency for smart materials such as photovoltaics and thermoelectrics is typically much less than for more conventional technologies, the potential utility of the energy is much greater. For example, the direct relationship between input energy and output energy renders many of the energy-exchanging smart materials, including piezoelectrics, pyroelectrics and photovoltaics, as excellent environmental sensors. The form of the output energy can further add direct actuation capabilities such as those currently demonstrated by electrostrictives, chemoluminescents and conducting polymers.

Reversibility/directionality

Many of the materials in the two above classes also exhibit the characteristic either of reversibility or of bi-directionality. Several of the electricity converting materials can reverse their input and output energy forms. For example, some piezoelectric materials can produce a current with an applied strain or can deform with an applied current. Materials with bi-directional property change or energy exchange behaviors can often allow further exploitation of their transient change rather than only of the input and output energies and/or properties. The energy absorption characteristics of phase-changing materials can be used either to stabilize an environment or to release energy to the environment depending on in which direction the phase change is taking place. The bi-directional nature of shape memory alloys can be exploited to produce multiple or switchable outputs, allowing the material to replace components comprised of many parts.

Size/location

Regardless of the class of smart material, one of the most fundamental characteristics that differentiate them from traditional materials is the discrete size and direct action of the material. The elimination or reduction in secondary transduction networks, additional components, and, in some cases, even packaging and power connections allows the minimization in size of the active part of the material. A component or element composed of a smart material will not only be much smaller than a similar construction using more traditional materials but will also require less infrastructural support. The resulting component can then be deployed in the most efficacious location. The smaller size coupled with the directness of the property change or energy exchange renders these materials to be particularly effective as sensors: they are less likely to interfere with the environment that they are measuring, and they are less likely to require calibration adjustments.

Type characterizations

For this discussion, we will distinguish between these two primary classes of smart materials discussed above by calling them Type 1 and Type 2 materials:

- Type 1 – a material that changes one of its properties (chemical, mechanical, optical, electrical, magnetic or thermal) in response to a change in the conditions of its environment and does so without the need of external control.

- Type 2 – a material or device that transforms energy from one form to another to effect a desired final state.

The note in Chapter 1 on the confusion of meanings of the term 'material' is particularly relevant here. Several of the descriptions given below for these smart material types edge into what are better described as products or devices since they either consist of multiple types of individual materials or they assume a product form. For example, electrochromism is a phenomenon, but electrochromic 'materials' invariably involve multiple layers of different materials serving specific functions that enable the phenomenon to be manifest. None the less, common usage by engineers and designers would typically broadly refer to an artifact of this type as a smart 'material', largely because of the way it is used in practice.

Figure 4–1 and the following two sections briefly describe the basic characteristics of a number of common Type 1 and Type 2 smart materials. There are, of course, many others.

TYPE OF SMART MATERIAL	INPUT		OUTPUT
Type 1 Property-changing			
Thermomochromics	Temperature difference		Color change
Photochromics	Radiation (Light)		Color change
Mechanochromics	Deformation		Color change
Chemochromics	Chemical concentration		Color change
Electrochromics	Electric potential difference		Color change
Liquid crystals	Electric potential difference		Color change
Suspended particle	Electric potential difference		Color change
Electrorheological	Electric potential difference		Stiffness/viscosity change
Magnetorheological	Electric potential difference		Stiffness/viscosity change
Type 2 Energy-exchanging			
Electroluminescents	Electric potential difference		Light
Photoluminescents	Radiation		Light
Chemoluminescents	Chemical concentration		Light
Thermoluminescents	Temperature difference		Light
Light-emitting diodes	Electric potential difference		Light
Photovoltaics	Radiation (Light)		Electric potential difference
Type 2 Energy-exchanging (reversible)			
Piezoelectric	Deformation	⟷	Electric potential difference
Pyroelectric	Temperature difference	⟷	Electric potential difference
Thermoelectric	Temperature difference	⟷	Electric potential difference
Electrorestrictive	Electric potential difference	⟷	Deformation
Magnetorestrictive	Magnetic field	⟷	Deformation

▲ **Figure 4-1** Sampling of different Type 1 and Type 2 smart materials in relation to input and output stimuli

Specific applications in design for these and other materials will be discussed in subsequent chapters.

4.2 Type 1 smart materials – property-changing

CHROMICS OR 'COLOR-CHANGING' SMART MATERIALS

Fundamental characteristics of chromics

A class of smart materials that are invariably fascinating to any designer is the so-called 'color-changing' material group which includes the following:

- *Photochromics* – materials that change color when exposed to light
- *Thermochromics* – materials that change color due to temperature changes.
- *Mechanochromics* – materials that change color due to imposed stresses and/or deformations.
- *Chemochromics* – materials that change color when exposed to specific chemical environments.
- *Electrochromics* – materials that change color when a voltage is applied. Related technologies include *liquid crystals* and *suspended particle* devices that change color or transparencies when electrically activated.

These constitute a class of materials in which a change in an external energy source produces a property change in the optical properties of a material – its absorptance, reflectance, or scattering. So-called 'color-changing' materials thus do not really change color. They change their optical properties under different external stimuli (e.g., heat, light or a chemical environment), which we often perceive as a color change. As was discussed previously, our perception of color depends on both external factors (light and the nature of the human eye) and internal factors such as those noted above. An understanding of these materials is thus more complicated than simply saying that they 'change colors'.

Recall that the external factors that affect our perception of color are many. Color is fundamentally a property of light. All incident light can be characterized by its spectral distribution of electromagnetic wavelengths. Surfaces can only reflect, absorb or transmit the available wavelengths – as such they are always subtractive. The human eye is also a subtractive surface, but does so comparatively. As a result, depending on the spectral and intensity distributions within the field of view, color is also *relative* within the context of the human eye.

Of direct interest herein is that the observed color of an object also depends on the intrinsic optical qualities of a material. In our discussion of fundamental material properties, we noted that atomic structures include negatively charged electrons. Since light consists fundamentally of energy impulses, it reacts with the negatively charged electrons in a material. Depending on the crystalline or molecular structure of the material, the light that attempts to pass through may be delayed, redirected, absorbed or converted to some other type of energy. The precise crystalline or molecular structure of the material present will determine which of these possible behaviors will take place, and in turn determine what wavelengths of light are in some way altered (which in turn affects the perceived color of the material). Interestingly, it is the molecular structure first encountered on a material's *surface* that determines the resultant behavior. As such, thin films, coatings and paints will predominantly determine the response to light, more so than the substrate.

In the case of a smart material with apparent color-changing properties, the intrinsic optical properties – absorptance, reflectance, scattering – of the material are designed to change with the input of external energy. Fundamentally, the input energy produces an altered molecular structure or orientation on the surface of the material on which light is incident. The structure depends on chemical composition as well as organization of the crystal or the molecule. This external energy can be in several forms (e.g., heat or radiant energy associated with light), but in each case it induces some change in the internal surface structures of the material by reacting with the negatively charged electrons present. These changes in turn affect the material's absorptance or reflectance characteristics and hence its perceived color. These changes can be over the entire spectrum or be spectrally selective. Interestingly, these changes are reversible. When the external energy stimulus disappears, an altered structure reverts back to its original state.

The main classes of color-changing smart materials are described by the nature of the input energy that causes the property change, and include *photochromics, electrochromics, thermochromics, mechanochromics,* and *chemochromics.*

Photochromic materials

Photochromic materials absorb radiant energy which causes a reversible change of a single chemical species between two

Naphthopyrans

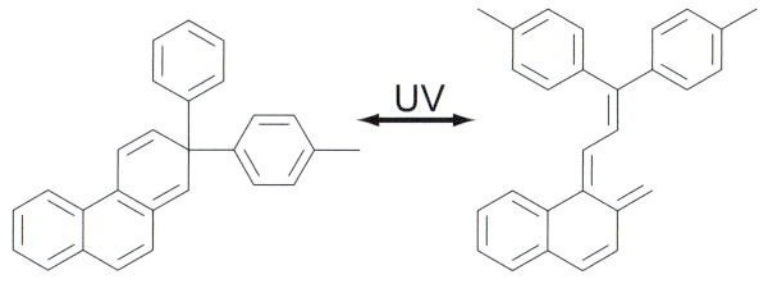

UV

The molecular structure changes (a twisting in this case) due to exposure to the input of radiant energy from light

▲ **Figure 4-2** Photochromic materials change color when exposed to light (a change in the molecular structure of a photochromic material causes a change in its optical properties)

different energy states, both of which have different absorption spectra. Photochromic materials absorb electromagnetic energy in the ultraviolet region to produce an intrinsic property change. Depending on the incident energy, the material switches between the reflectively and absorptively selective parts of the visible spectrum. The molecule used for photochromic dyes appears colorless in its unactivated form. When exposed to photons of a particular wavelength, the molecular structure is altered into an excited state, and thus it begins to reflect at longer wavelengths in the visible spectrum. On removal of the ultraviolet (UV) source, the molecule will revert to its original state. A typical photochromic film, for example, can be essentially transparent and colorless until it is exposed to sunlight, when the film begins selectively to reflect or transmit certain wavelengths (such as a transparent blue). Its intensity depends upon the directness of exposure. It reverts to its original colorless state in the dark when there is no sunlight.

Photochromic materials are used in a wide range of applications. Certainly we see them used in a wide range of consumer products, such as sunglasses that change their color. In architecture, they have been used in various window or façade treatments, albeit with varying amounts of success,

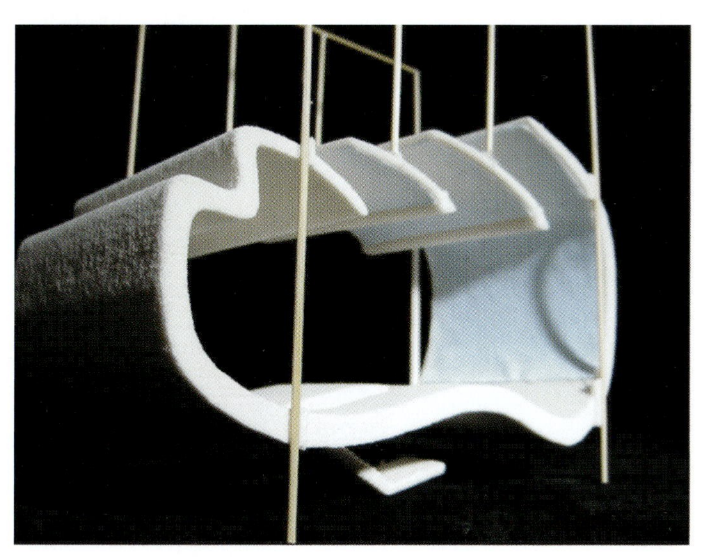

▲ **Figure 4-3** Design Experiment: In the proposed 'Coolhouse', interior panels are covered with photochromic cloth that changes from a base color of white to blue upon exposure to sunlight. The panel shapes are designed for a particular solar angle for a specified time and place during the summer. At this time, the interior becomes a cool blue. In the winter, the cloth is not exposed and the interior remains white. (Teran and Teman Evans)

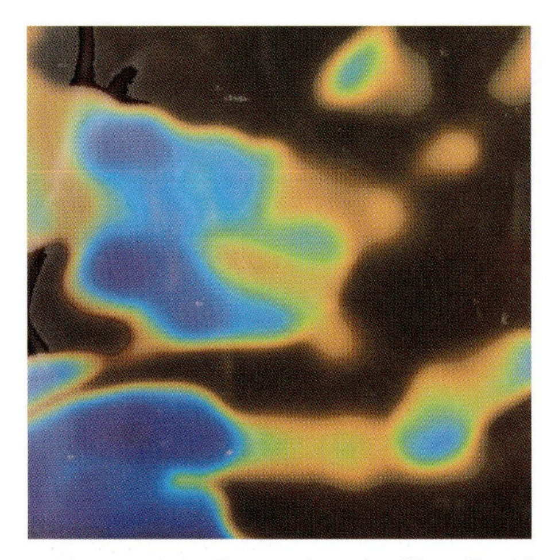

▲ **Figure 4-4** Thermochromic film (liquid crystal) calibrated for 25–30° C. Different colors indicate different temperature levels in the film. Blue is the highest temperature level and black is the lowest

to control solar gain and reduce glare. By and large, these applications have not proven effective because of the slowness of response and heat gain problems. Chapters 6 and 7 will treat these applications in greater detail.

Thermochromic materials

Thermochromic materials absorb heat, which leads to a thermally induced chemical reaction or phase transformation. They have properties that undergo reversible changes when the surrounding temperature is changed. The liquid crystal film versions can be formulated to change temperature from −25 to +250 °F (−30 to 120 °C) and can be sensitive enough to detect changes as small as 0.2 °F.

Thermochromic materials come in many forms, including liquid crystal forms used in thermochromic films and the leucodyes used in many other applications. Films are used in applications such as battery testers, thermometers and so forth. The widely used 'band thermometer' that is placed on a person's forehead, for example, is made of thermochromic materials designed to be sensitive to particular temperature levels. A simple visual calibration device signifies the temperature level corresponding to a particular color. They can be precisely calibrated. Leucodyes, by contrast, are used in various paints and papers.

In architecture and furniture design, the seemingly never-ending quest to show the past presence of a person at a particular location or on a piece of furniture has found a new tool for expression. Several of Jurgen Mayer H.'s furniture and

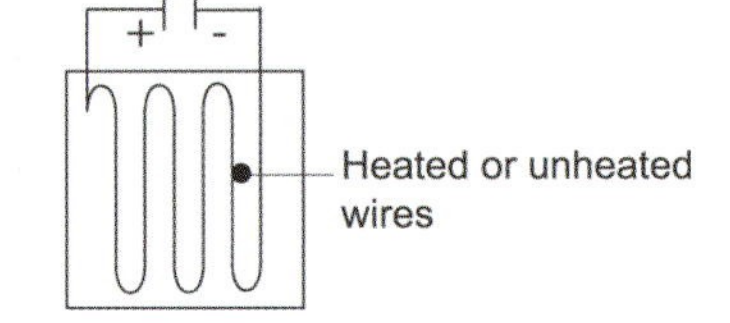

▲ **Figure 4-5** Design experiment: in this simple setup, heated wires are used to generate a specific color change pattern on a thermochromic material. (Antonio Garcia Orozco)

▲ **Figure 4-6** Memories of touch via thermochromic materials. (Courtesy Juergen Mayer H)

consumer goods pieces, for example, are sensitive to body heat and show a colored 'imprint' of a person who just sat on the furniture. The image fades with time.

The notion of using thermochromic materials on the exterior of a building has similarly always aroused interest. Unfortunately, a major problem with the use of currently available thermochromic paints on the exterior is that exposure to ultraviolet wavelengths in the sun's light may cause the material to degrade and lose its color-changing capabilities.

Mechanochromic and chemochromic materials

Mechanochromics have altered optical properties when the material is subjected to stresses and deformations associated with external forces. Many polymers have been designed to exhibit these kinds of properties. The old household device for imprinting raised text onto plastic strips utilizes a plastic of this type. The raised text that results from a mechanical deformation shows through as a different color.

Chemochromics include a wide range of materials whose properties are sensitive to different chemical environments. You might perhaps recall the ancient litmus paper in a basic chemistry class.

Electrochromic materials

Electrochromism is broadly defined as a reversible color change of a material caused by application of an electric current or potential. An electrochromic window, for example, darkens or lightens electronically. A small voltage causes the glazing material to darken, and reversing the voltage causes it to lighten.

There are three main classes of materials that change color when electrically activated: electrochromics, liquid crystals and suspended particles. These technologies are not one-constituent materials, but consist of multi-layer assemblies of different materials working together.

Fundamentally, color change in an electrochromic material results from a chemically induced molecular change at the surface of the material through oxidation-reduction. In order to achieve this result, layers of different materials serving different ends are used. Briefly, hydrogen or lithium ions are transported from an ion storage layer through an ion conducting layer, and injected into an electrochromic layer. In glass assemblies, the electrochromic layer is often tungsten oxide (WO_3). Applying a voltage drives the hydrogen or lithium ions from the storage layer through the conducting layer, and into the electrochromic layer, thus changing the optical properties of the electrochromic layer and causing it to absorb certain visible light wavelengths. In this case, the glass darkens. Reversing the voltage drives ions out of the electrochromic layer in the opposite direction (through the conducting layer into the storage layer), thus causing the glass to lighten. The process is relatively slow and requires a constant current.

The layers forming the electrochromic component can be quite thin and readily sandwiched between traditional glazing materials. Many companies have been developing products that incorporate these features in systems from as small as a residential window to as large as the curtain wall of a building. In a typical application, the relative transparency and color tint of electrochromic windows can be electrically controlled. Note, however, that it is necessary for the voltage to remain on for the window to remain in a darkened state. This can be disadvantageous for many reasons. In Chapters 6 and 7 we will return to a discussion of the applications of electrochromic technologies.

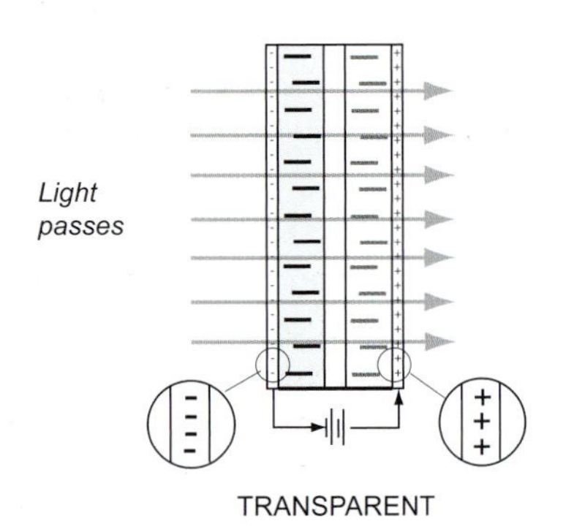

▲ **Figure 4-7** Electrochromic glass

PHASE-CHANGING MATERIALS

As discussed in the earlier section on phase changes in materials, many materials can exist in several different physical states – gas, liquid or solid – that are known as phases. A change in the temperature or pressure on a material can cause it to change from one state to another, thereby undergoing what is termed a 'phase change'. Phase change processes invariably involve the absorbing, storing or releasing of large amounts of energy in the form of latent heat. A phase change from a solid to a liquid, or liquid to a gas, and vice versa, occurs at precise temperatures. Thus, where energy is absorbed or released can be predicted based on the composition of the material. Phase-changing materials deliberately seek to take advantage of these absorption/release actions.

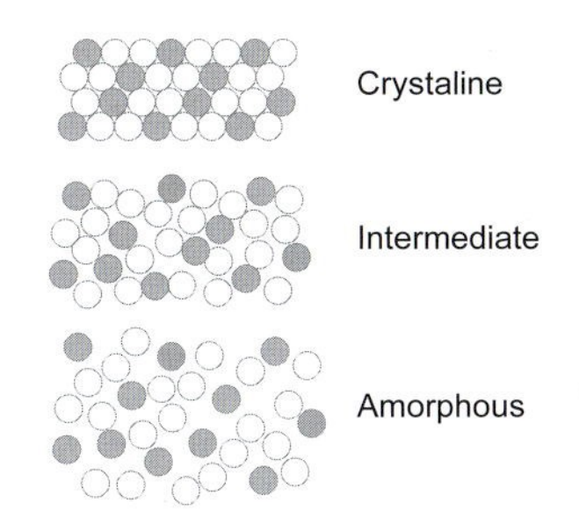

▲ **Figure 4-8** Phase change transformation

While most materials undergo phase changes, there are several particular compositions, such as inorganic hydrated salts, that absorb and release large amounts of heat energy. As the material changes from a solid to a liquid state, and then subsequently to a gaseous state, large amounts of energy must be absorbed. When the material reverts from a gaseous to a liquid state, and then to a solid state, large amounts of energy will be released. These processes are reversible and phase-changing materials can undergo an unlimited number of cycles without degradation.

Since phase-changing materials can be designed to absorb or release energy at predictable temperatures, they have naturally been explored for use in architecture as a way of helping deal with the thermal environment in a building. One early application was the development of so-called 'phase change wallboard' which relied on different embedded materials to impart phase change capabilities. Salt hydrates, paraffins and fatty acids were commonly used. The paraffin and fatty acids were incorporated into the wallboard initially by direct immersion. Subsequently, filled plastic pellets were used. Transition temperatures were designed to be around 65–72 °F for heating dominated climates with primary heating needs and 72–79 °F for climates with primary cooling needs.

Products based on direct immersion technologies never worked well and proved to have problems of their own that were associated with the more or less exposed paraffin and fatty acids (including problems with animals eating the wallboard products). Technologies based on sealing phase-changing materials into small pellets worked better. Pellet technologies have achieved widespread use, for example, in connection with radiant floor heating systems. In many climates, radiant floor systems installed in concrete slabs can provide quite comfortable heating, but are subjected to undesired cycling and temperature swings because of the need to keep the temperature of the slab at the desired level, which typically requires a high initial temperature. Embedding phase-changing materials in the form of encased pellets can help level out these undesirable temperature swings.

Phase-changing materials have also successfully found their way into outdoor clothing. Patented technologies exist for embedding microencapsulated phase-changing materials in a textile. These encapsulations are microscopic in size. The phase-changing materials within these capsules are designed to be in a half-solid, half-liquid state near normal skin temperature. As a person exercises and generates heat, the materials undergo a phase change and absorb excess heat, thus keeping the body cooler. As the body cools down, and heat is needed, the phase-changing materials begin to release heat to warm the body.

Of particular interest in the applications discussed is that successful applications of phase-changing materials occurred when they were encapsulated in one form or another. It is easy to imagine how encapsulated phase-changing materials could be used in many other products, from lamps to furniture, as a way of mitigating temperature swings.

CONDUCTING POLYMERS AND OTHER SMART CONDUCTORS

In this day and age of electronic devices, it is no wonder that a lot of attention has been paid to materials that conduct electricity. Any reader of scientific news has heard about the strong interest in materials such as superconductors that offer little or no resistance to the flow of electricity. In this section, however, we will look at a broader range of conducting materials, including those that offer great potential in different design applications.

In general, there is a broad spectrum associated with electrical conductivity through terms like 'insulators', 'conductors', 'semi-conductors' and 'super-conductors' – with insulators being the least conductive of all materials. Many of the products that architectural and industrial designers are most familiar with are simple conductors. Obviously, many metals are inherently electrically conductive due to their atomic bonding structures with their loosely bound electrons allowing easy electron flow through the material. As discussed in more detail in Chapter 6, many traditional products that are not intrinsically conductive, e.g., glasses or many polymers, can be made so by various means. Polymers can be made conductive by the direct addition of conductive materials (e.g., graphite, metal oxide particles) into the material. Glasses, normally highly insulating, can be made conductive and still be transparent via thin film metal deposition processes on their surfaces.

There are other polymers whose electrical conductivity is intrinsic. Electroactive polymers change their electrical conductivity in response to a change in the strength of an electrical field applied to the material. A molecular rearrangement occurs, which aligns molecules in a particular way and frees electrons to serve as electricity conductors. Examples include polyaniline and polypyrrole. These are normally conjugated polymers based on organic compounds that have internal structures in which electrons can move more freely. Some polymers exhibit semiconductor behavior and can be light-emitting (see *Semiconductors* below and *Light-emitting polymers* in Chapter 6). Electrochemical polymers exhibit a change in response to the strength of the chemical environment present.

A number of applications have been proposed for conducting polymers. Artificial muscles have been developed using polypyrrole and polyaniline films. These films are laminated around an ion-conducting film to form a sandwich construction. When subjected to a current, a transfer of ions occurs. The current flow tends to reduce one side and oxidize the other. One side expands and the other contracts. Since the films are separated, bending occurs. This bending can then be utilized to create mechanical forces and actions.

Despite the dream of many designers to cover a building with conducting polymers, and to have computer-generated images appearing anywhere one desires, it is necessary to remember that these materials are essentially conductors only. In the same way it would not be easy to make an image appear on a sheet of copper, it is similarly difficult to make an image appear on a conducting polymer. Since films can be manipulated (cut, patterned, laminated, etc.), possibilities in this realm do exist, but remain elusive.

Other smart conductors include *photoconductors* and *photoresistors* that exhibit changes in their electrical conductivity when exposed to a light source. *Pyroconductors* are materials whose conductivities are temperature-dependent, and can have minimal conductivity near certain critical low temperatures. *Magnetoconductors* have conductivities responsive to the strength of an applied magnetic field. Many of these specialized conducting materials find applications as sensors of one type or another. Many small devices, including motion sensors, already employ various kinds of photoconductors or photoresistors (see Chapter 7). Others, including pyroconductors, are used for thermal sensing.

RHEOLOGICAL PROPERTY-CHANGING MATERIALS

The term 'rheological' generally refers to the properties of flowing matter, notably fluids and viscous materials. While not among the more obvious materials that the typical designer would seek to use, there are many interesting properties, in particular viscosity, that might well be worth exploring.

Many of these materials are termed 'field-dependent'. Specifically, they change their properties in response to electric or magnetic fields. Most of these fluids are so-called 'structured fluids'' with colloidal dispersions that change phase when subjected to an electric or magnetic field. Accompanying the phase change is a change in the properties of the fluid.

Electrorheological (ER) fluids are particularly interesting. When an external electric field is applied to an electrorheological fluid, the viscosity of the fluid increases remarkably. When the electric field is removed, the viscosity of the fluid reverts to its original state. Magnetorheological fluids behave similarly in response to a magnetic field.

The changes in viscosity when electrorheological or magnetorheological fluids are exposed to electric or magnetic fields, respectively, can be startling. A liquid is seemingly transformed into a solid, and back again to a liquid as the field is turned off and on.

These phenomena are beginning to be utilized in a number of products. An electrorheological fluid embedded in an automobile tire, for example, can cause the stiffness of the tire to change upon demand; thus making it possible to 'tune' tires for better cornering or more comfortable straight riding. Some devices that typically require mechanical interfaces, e.g., clutches, might conceivably use smart rheological fluids as replacements for mechanical parts.

In architecture and industrial design, little use has been made of smart rheological fluids. One can imagine, however, chairs with smart rheological fluids embedded in seats and arms so that the relative hardness or softness of the seat could be electrically adjusted. The same is obviously true for beds.

LIQUID CRYSTAL TECHNOLOGIES

Liquid crystal displays are now ubiquitously used in a host of products. It would be hard to find someone in today's modern society that has not seen or used one. This widespread usage, however, does not mean that liquid crystal technologies are unsophisticated. Quite the contrary; they are a great success story in technological progress.

Liquid crystals are an intermediate phase between crystalline solids and isotropic liquids. They are orientationally ordered liquids with anisotropic properties that are sensitive to electrical fields, and therefore are particularly applicable for optical displays. Liquid crystal displays utilize two sheets of polarizing material with a liquid crystal solution between them. An electric current passed through the liquid causes the crystals to align so that light cannot pass through them. Each crystal is like a shutter, either allowing light to pass through or blocking the light.

▲ **Figure 4-9** Progressive phase change of nematic liquid crystal films (the typical thermotropic liquid crystal similar to what is used in LCDs). (Images courtesy of Oleg D. Lavrentovich of the Liquid Crystal Institute, Kent State University)

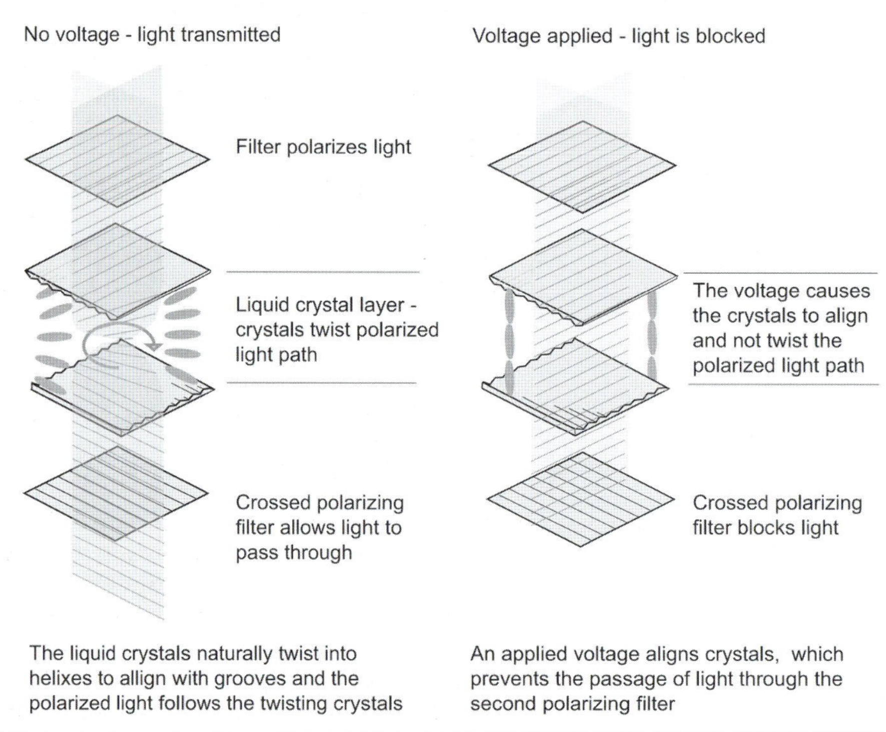

▲ **Figure 4-10** A liquid crystal display (LCD) uses two sheets of polarizing material and a liquid crystal solution sandwiched in between them

SUSPENDED PARTICLE DISPLAYS

Newly developed suspended particle displays are attracting a lot of attention for both display systems and for more general uses. These displays are electrically activated and can change from an opaque to a clear color instantly and vice-versa. A typical suspended particle device consists of multiple layers of different materials. The active layer associated with color change has needle-shaped particles suspended in a liquid. (films have also been used). This active layer is sandwiched between two parallel conducting sheets. When no voltage is applied, the particles are randomly positioned and absorb light. An applied voltage causes the particles to align with the field. When aligned, light transmission is greatly increased through the composite layers.

Interestingly, the color or transparency level remains at the last setting when voltage was applied or turned off. A constant voltage need not be applied for the state to remain.

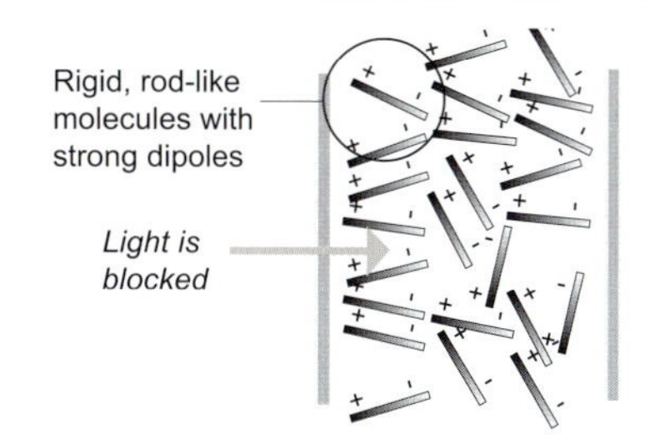

Particles suspended in film between two clear conducting layers align randomly in the absence of an electric field, absorbing light

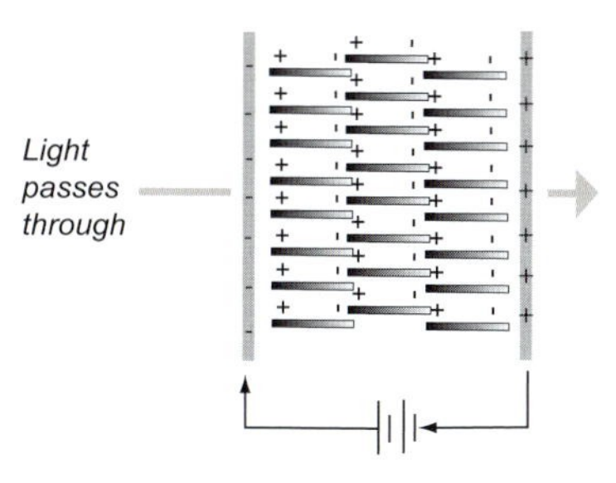

Application of an electric field causes individual molecules to orient similarly, thus allowing light to pass through

▲ **Figure 4-11** Suspended particle display

Long chain molecular structures

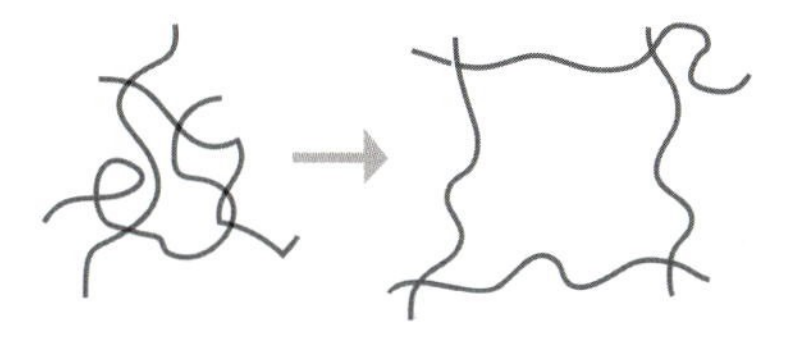

Large reversible volume changes can occur due to changes in the surrounding environment.

Polyacrylamide polymer crystals with a strong affinity for water swell to several hundred times their size in water, and then can revert back to their original size on drying

▲ **Figure 4-12** Volume-changing polymer gels

OTHER TYPE I MATERIALS

There are a great many other interesting materials that exhibit one form or another of property change. Shape-changing gels or crystals, for example, have the capacity to absorb huge amounts of water and in doing so increase their volumes by hundred-folds. Upon drying out, these same materials revert to their original sizes (albeit often in a deformed way). Applications are found in everything from dehumidification devices and packaging through to baby diapers and plant watering spikes.

4.3 Type 2 smart materials – energy-exchanging

Energy fields – environments – surround all materials. When the energy state of a given material is equivalent to the energy state of its surrounding environment, then that material is said to be in equilibrium: no energy can be exchanged. If the material is at a different energy state, then a potential is set up which drives an energy exchange. All of the energy exchange materials involve atomic energy levels – the input energy raises the level, the output energy returns the level to its ground state. For example, when solar radiation strikes a photovoltaic material, the photon energy is absorbed, or more precisely – absorbed by the atoms of the material. As energy must be conserved, the excess energy in the atoms forces the atom to move to a higher energy level. Unable to sustain this level, the atom must release a corresponding amount of energy. By using semi-conductor materials, photovoltaics are able to capture this release of energy – thereby producing electricity. Note that all materials – traditional as well as smart – must conserve energy, and as such the energy level of the material will increase whenever energy is input or added. For most materials, however, this increase in energy manifests itself by increasing the internal energy of the material, most often in the form of heat. Energy exchange smart materials distinguish themselves in their ability to recover this internal energy in a more usable form.

Many of the energy-exchanging materials are also bi-directional – the input energy and output energy can be switched. The major exceptions to this are materials that exchange radiation energy – the high inefficiency of radiant energy exchange increases thermodynamic irreversibility. Furthermore, unlike most (although not all) of the property-changing materials, the energy-exchange materials are almost always composite materials – exceptions include magnetostrictive iron and naturally occurring piezoelectric quartz.

The following sections describe a number of commonly used Type 2 energy-exchanging materials.

▲ **Figure 4-13** Three types of fluorescing calcite crystals (middle image also has fluorite mixed in) (Images courtesy of Tema Hecht and Maureen Verbeek)

LIGHT-EMITTING MATERIALS

Luminescence, fluorescence and phosphorescence

A definition of *luminescence* can be backed into by saying that it is emitted light that is not caused by incandescence,[1] but rather by some other means, such as chemical action. More precisely, the term luminescence generally refers to the emission of light due to incident energy. The light is caused by the re-emission of energy in wavelengths in the visible spectrum and is associated with the reversion of electrons from a higher energy state to a lower energy state. The phenomenon can be caused by a variety of excitation sources, including electrical, chemical reactions, or even friction. A classic example of a material that is luminescent due to a chemical action is the well-known 'light stick' used for emergency lighting or by children during Halloween.

Luminescence is the general term used to describe different phenomena based on emitted light. If the emission occurs more or less instantaneously, the term *fluorescent* is used. Fluorescents glow particularly brightly when bathed in a 'black light' (a light in the ultraviolet spectrum). If the

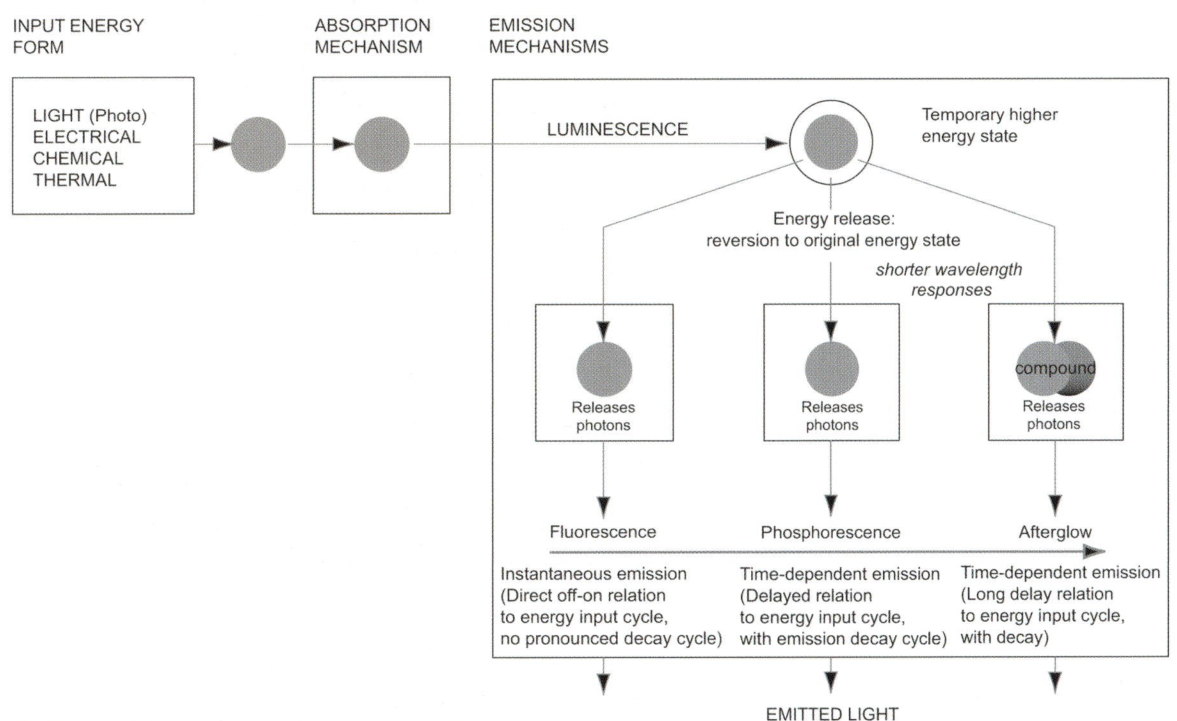

▲ **Figure 4-14** Diagram showing general phenomenon of luminescence

emission is slower or delayed to several microseconds or milliseconds, the term *phosphorescence* is used. Many compounds are either naturally phosphorescent or designed to be so. The amount of delay time depends on the particular kind of phosphor used. Common phosphors include different metal sulfides (e.g., ZnS). Common television screens rely on the use of ZnS. Strontium aluminate is also strongly phosphorescent. In some situations, the light emission can continue long after the source of excitation is removed because the electrons become temporarily trapped because of material characteristics. Here the term *afterglow* is used.

Most materials that are luminescent are solids that contain small impurities, e.g., zinc sulfates with tiny amounts of copper. When these materials are exposed to incident energy in any of several forms, the energy associated with the impinging electrons or photons is absorbed by the material, which in turn causes electrons within the material to rise to a higher level. Following the descriptive model suggested by Flinn and Trojan, these electrons subsequently may fall into what are commonly called 'traps' associated with the impurities.[2] After a while, a trapped electron gains enough energy to leave its trap and in doing so produces a light photon in a wavelength in the visible spectrum. Its wavelength is dependent on the ion (e.g., copper) producing the trap. Thus, the nature of the emitted light and its speed and duration of emission depend upon the type of impurity present.

Different properties, including the color of the emitted light, can be engineered by varying different compounds and impurity inclusions to yield specific kinds of light-emitting materials. In particular, there is a constant quest to improve the duration of the phosphorescent effect once the excitation source has been removed. Materials such as strontium aluminate, for example, have been exploited for use because of their long afterglow duration once the excitation source has been removed.

Photoluminescence generally refers to a kind of luminescence that occurs when incident energy associated with an external light source acts upon a material that then re-emits light at a lower energy level. A process of electronic excitation by photon absorption is involved. As a consequence of energy conservation, the wavelength of the emitted light is longer (i.e., 'redder' and involves less energy) than the wavelength of the incident light. Several kinds of phosphors photoluminesce brightly, particularly when exposed to ultraviolet light.

Typical fluorescent lamps are also based on photoluminescent effects. The inside of a lamp is coated with a phosphor that is excited by ultraviolet mercury radiation from a glow discharge.

In *chemoluminescence*, the excitation comes from a chemical action of one type or another. The lightstick mentioned earlier still provides the best common example of this phenomenon. Particularly interesting here is that chemoluminescence produces light without a corresponding heat output, which is surprising since a chemical reaction is involved. If the temperature of the surrounding heat environment is increased, however, there will be an increase in the reaction time, hence light output, and a reduction in temperature will correspondingly reduce the light output.

A subset of chemoluminescence normally called *bioluminescence* is particularly fascinating because it provides the glow associated with various light-emitting insects, such as fireflies, or fish such as the Malacosteus, which navigates the depths of the sea via its own night light. Consider, for example, the squid that can alter its luminescence to match either moonlight or sunlight.

Electroluminescence

With *electroluminescent* materials the source of excitation is an applied voltage or an electric field. The voltage provides the energy required. There are actually two different ways that electroluminescence can occur. The first and typical condition occurs when there are impurities scattered through the basic phosphor. A high electric field causes electrons to move through the phosphor and hit the impurities. Jumps occurring in connection with the ionized impurity cause luminescence to occur. The color emitted is dependent on the type of impurity material that forms the active ions. A second and more complex behavior occurs in special materials, such as semiconductors, because of a general movement of electrons and holes (see *Semiconductors*, *Lasers* and *LEDS (light-emitting diodes)* below).

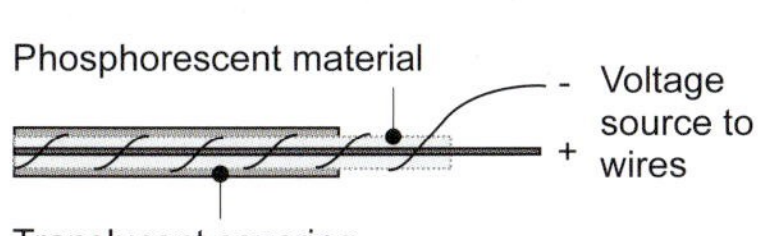

▲ **Figure 4-15** Electroluminescent wire

Electroluminescent materials are widely used for light strips and panels of all descriptions. The bright backlights in inexpensive watches are invariably electroluminescent panels. As noted above, colors are dependent on the active ions selected for use. In very inexpensive systems, however, simple colored filters are used to give variety. Strips or panels can be designed to work off of different applied voltages. They can be battery operated. On the other hand, larger panels can be made to respond to household voltages.

Since the luminescent effect depends on phosphors and an electric field, electroluminescent strips or panels can be made

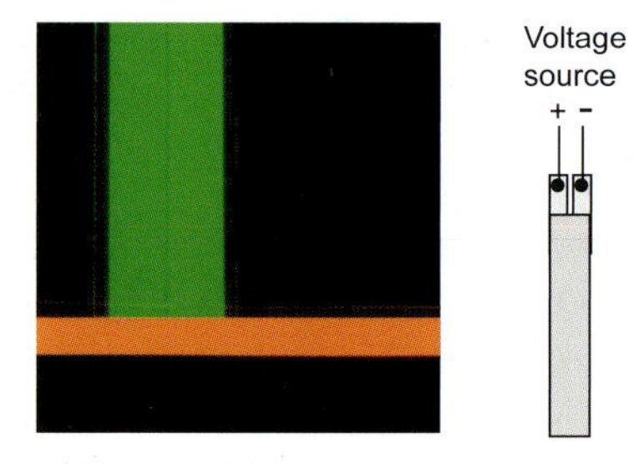

▲ **Figure 4-16** Electroluminescent strips

using a variety of different neutral substrates. Very simple strips can be made in which a phosphorous material is applied evenly to a polymeric strip, and covered by another transparent strip for protection. A small wire to provide the electric field is applied to the strip. Voltage sources can be batteries. Larger panels can be made using polymeric materials as well. An electric strip would essentially surround the panel. Interestingly, these polymeric panels can literally be cut into different shapes as long as the electrical field can be maintained. Other materials that are often used as substrates include glass, ceramics and plastics.

Electroluminescent lamps are becoming widely used. They draw little power and generate no heat. They provide a uniformly illuminated surface that appears equally bright from all angles. Since they do not have moving or delicate parts, they do not break easily. Chapter 6 discusses applications in more detail.

BASIC SEMICONDUCTOR PHENOMENA

Few people have not heard of *semiconductors* – the materials that have helped usher in an age of high-powered microelectronic devices. The basic phenomenon underlying a semiconductor forms the basis for other technologies as well, including transistors and, of special interest in this book, the photovoltaic effect associated with solar power. Few people, of course, have any idea of what this phenomenon actually involves and what semiconducting devices actually do. Here we will only touch on the salient features of these complex materials.

Basic semiconductor materials, such as silicon, are neither good conductors nor good insulators, but, with the addition of small impurities called *dopants*, they can be made to possess many fascinating electrical properties. The addition of these dopants or impurities allows electron movements to be precisely controlled. Exploitation of the resultant properties has allowed a semiconductor to serve the same functions as complicated multipart electronic circuitries.

Silicon is the most widely used semiconducting material, although other material types are possible. Basic semiconducting materials exhibit interesting properties when surrounding temperatures are varied. Unlike most metals wherein increases in temperatures cause increases in resistance, the conductivity of semiconducting materials increases with increasing temperatures. This property already makes it quite attractive for many applications. It results from a particular type of electron band structure in the internal structure of the materials. A gap exists between bands through which thermally excited electrons cross in particular conditions.

The addition of dopants or impurities creates other conditions. The role of impurities with respect to light-emitting materials was previously noted. Of importance in this discussion is the role of the impurities in affecting the flow of electrons through a material. Here again the flow is affected, but in this case in a controllable way. Silicon matrix materials are alloyed with specific concentrations of a dopant, such as boron, via a complex deposition layering procedure to form a semiconductive device. Multiple dopants of different types may be used. The specific nature of these assemblies determines their useful electronic properties.

Figure 4–17 illustrates a typical makeup of a device that consists of a junction of so-called *p* and *n* semiconductor materials (made by using different dopants on silicon substrates). In the first type of material, *n*, electrons with a negative charge are predominantly present. In the second type, *p*, holes (locations of missing electrons) are primarily present resulting in a positive charge. Application of a negative charge to the *p* side causes the charges to be electrostatically attracted away from each other, creating a zone that is free of electrons. No current flows through this region. Application of a positive charge to the *p* side causes the reverse situation. Electrons flow through the barrier zone creating a current.

Semiconductor materials with different added selected impurities (*dopants*) - conductivities increase with temperatures in semiconductor materials

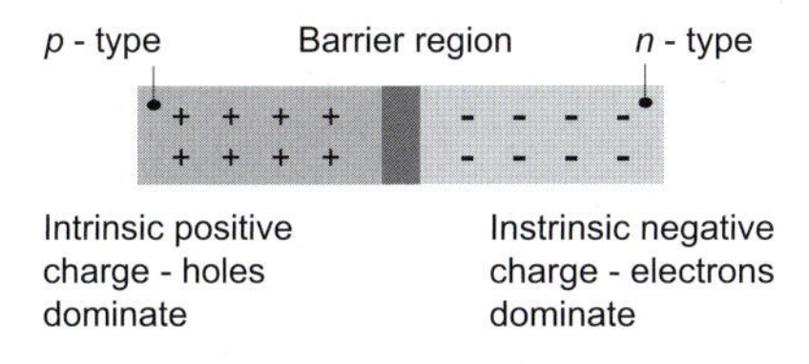

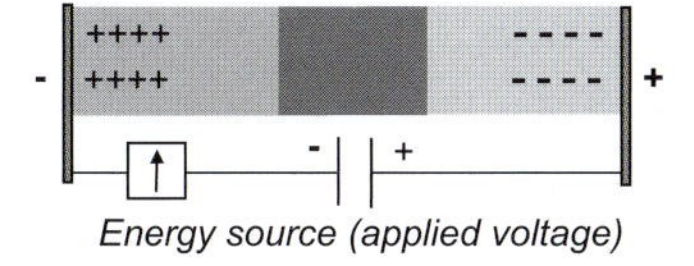

In the *reverse bias* mode, there is no flow of current across the barrier region

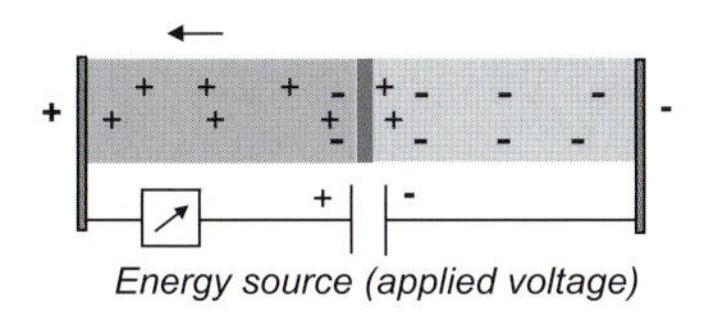

In the *forward bias* mode, the current increases exponentially with the applied voltage

▲ **Figure 4-17** Basic semiconductor behavior

PHOTOVOLTAICS, LEDS, TRANSISTORS, THERMOELECTRICS

Many widely used devices have their fundamental basis in semiconductor technology. *Photovoltaic* technologies are discussed in detail elsewhere (see Chapter 7). Here it is important to note that the basic underlying phenomenon is related to the semiconductor behavior noted above. A photovoltaic device consists primarily of a *p* and *n* junction. Instead of there being an applied voltage as described above, however, there is an incident energy (typically solar) that acts on the junction and provides the external energy input. In typical solar cells, the *n* layer is formed on top of the *p* layer. Incident energy impinges on the *n* layer. This incident energy causes a change in electron levels that in turn causes adjacent electrons to move because of electrostatic forces. This movement of electrons produces a current flow. Phototransistors are similar in that they convert radiant energy from light into a current.

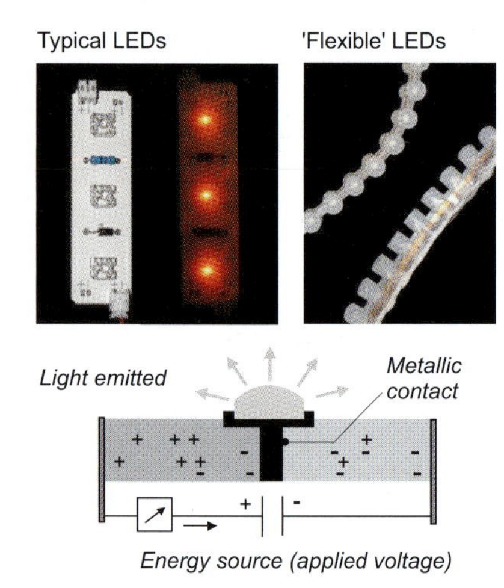

In a *light-emitting diode (LED)*, energy input into the junction creates a voltage output

▲ **Figure 4-18** Light emitting diodes (LED) are based on semiconductor technologies

Energy source - radiant energy from light

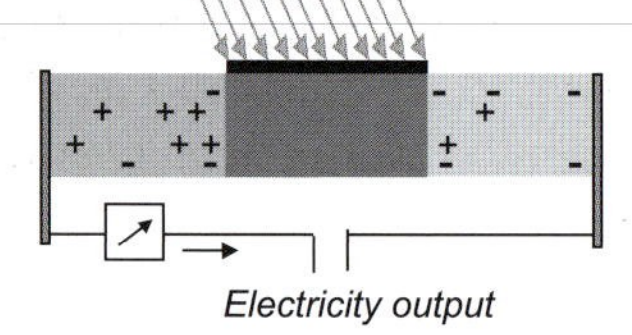

a *photovoltaic device (solar cell)*, energy put into the junction creates a voltage output

▲ **Figure 4-19** Photovoltaic (PV) devices are based on semiconductor technologies

Common *LEDs (light-emitting diodes)* are based essentially on the converse of photovoltaic effects. An LED is a semiconductor that luminesces when a current passes through it. It is basically the opposite of a photovoltaic cell. LEDs are discussed in detail in Chapter 7. Transistors are similarly based on semiconductor technologies. Fundamentally, a transistor can be used as a signal amplification device, or as a switching device.

Thermoelectrics or *Peltier* devices are an electronic form of heat pump. A typical Peltier device uses a voltage input to create hot and cold junctions, hence they can be used for heating or cooling. They are found in computers as cooling devices, and in common automotive and household goods as small heaters or coolers. When in use, there must be a way provided to carry the heat generated away from the unit. In larger units, fans are commonly used.

Lasers are one of the ubiquitous workhorses of today's technological society. Laser light occurs via stimulated emission. In a laser, an electron can be caused to move from one energy state to another because of an energy input, and, as a consequence, emit a light photon. This emitted photon can in turn stimulate another electron to change energy levels and emit another photon that vibrates in phase with the first. The chain builds up quickly with increasing intensity. Emitted photons vibrate in phase with one another. Hence the light is phase-coherent. The term 'coherent light' is often used. The light is monochromatic, which in turn allows it to be highly focused. Since the light occurs via stimulated emissions, the acronym Laser was adopted (i.e., light amplification by stimulated emission of radiation).

Many types of lasers exist that rely on different methods of excitation and use different materials. There are ruby lasers,

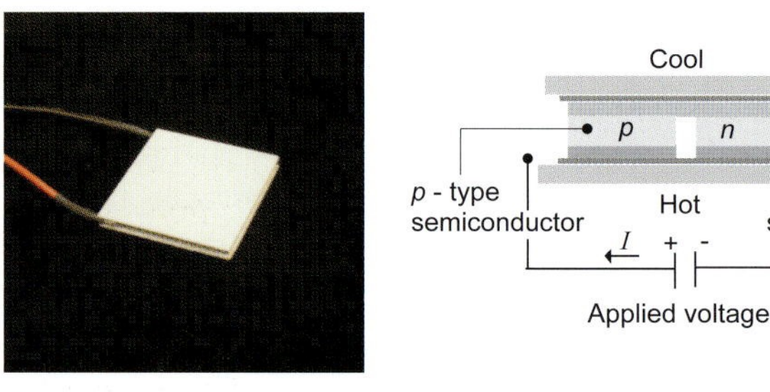

▲ **Figure 4-20** In a Peltier device an input electrical current causes one face to heat up and the other to cool down. Ceramic plates are used in the device shown. It is necessary to transfer heat away from the hot surface via fans or heat spreaders. Peltier devices are used in many products, including drink coolers and in computers to cool microchips

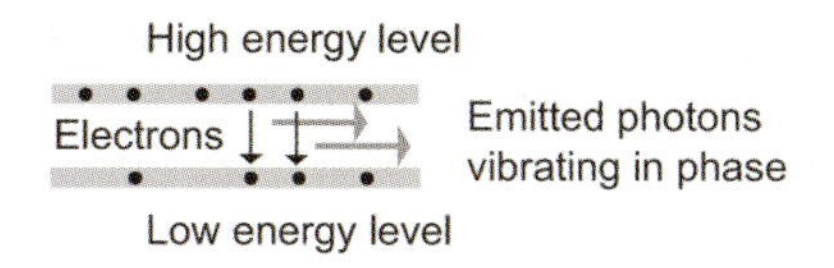

Basic principles - light amplification by stimulated emission of radiation (LASER)

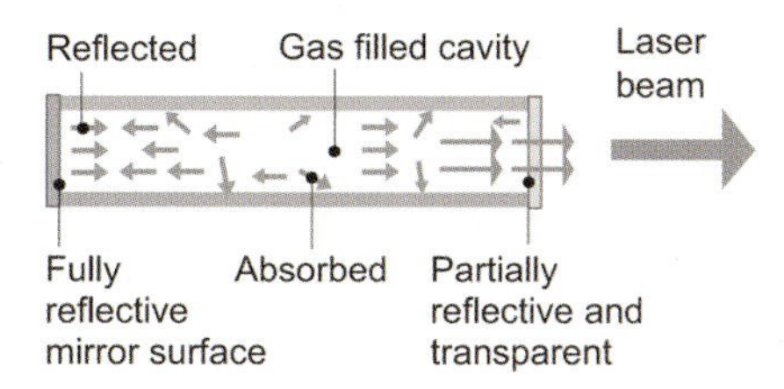

Gas lasers - photon are bounced back and forth with increasing intensity to create a high powered laser

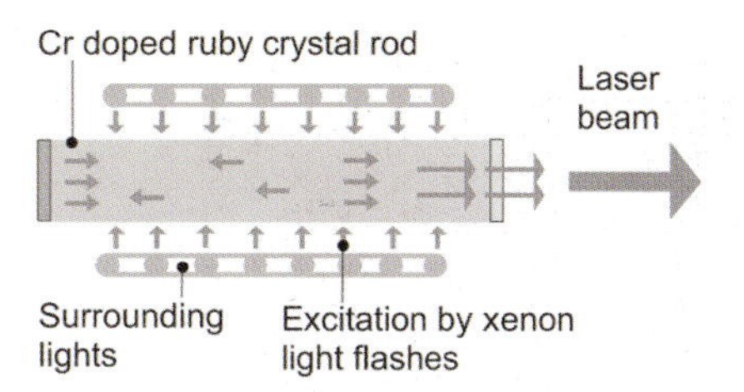

Ruby crystal laser - photons are excited by flashes from surrounding xenon lights

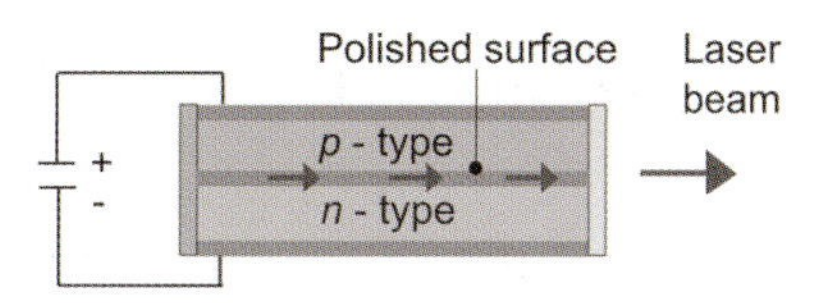

'Semiconductor' laser - commonly used in laser guides, laser printers, and surveying equipment

▲ **Figure 4-21** Lasers – basic principles and types

gas lasers and so forth. Powers can vary. Gas lasers can be quite powerful and cut many materials. The most ubiquitous kind of lasers used in printers, pointers, construction levels, surveying instruments etc. are typically based on semiconductor technologies (see Figure 4–21).

PIEZOELECTRIC EFFECTS AND MATERIALS

In this section we enter into the world of the *piezoelectric effect* that forms the underlying basis for products as diverse as some types of microphones and speakers, charcoal grill fire starters, vibration reducing skis, doorbell pushers and an endless number of position sensors and small actuators. All of these devices involve use of a piezoelectric material in which an applied mechanical force produces a deformation that in turn produces an electric voltage, or, conversely, an applied voltage that causes a mechanical deformation in the material that can be used to produce a force. This general phenomenon is called the piezoelectric effect.

The *piezoelectric phenomenon* (piezo means pressure in Greek) was observed by the brothers Pierre and Jacques Curie when they were 21 and 24 years old in 1880. They observed that when a pressure is applied to a polarized crystal, the mechanical deformation induced resulted in an electrical

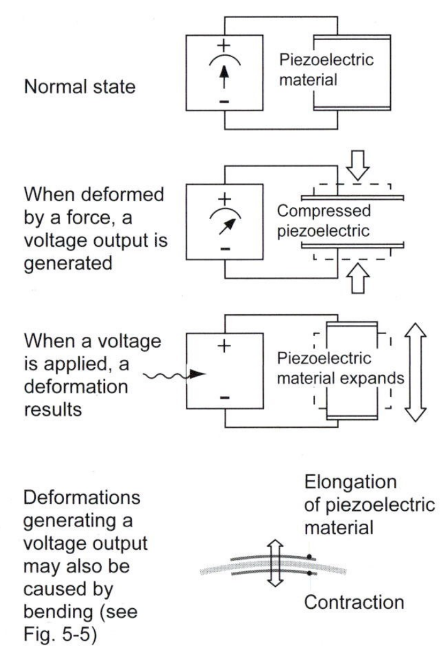

▲ **Figure 4-22** Piezoelectric behavior

charge. The phenomenon is based upon a reversible energy conversion between electrical and mechanical forms that occurs naturally in permanently polarized materials in which parts of molecules are positively charged and other parts are negatively charged. Many naturally found crystals (e.g., quartz) possess this property, as do many newly developed polymers and ceramics. The property is curiously similar to that found in magnets where permanent magnetic polarization occurs, except here we are dealing with electrical charges.

In piezoelectric materials, each cell or molecule is a dipole with a positive and negative charges onto either end. There is an alignment of the internal electric dipoles. This alignment can result in a surface charge, but this charge is neutralized by free charges present in the surrounding atmosphere. A force is applied to the piezoelectric material that causes deformations to take place, which in turn alters the neutralized state of the surface by changing the orientation of the dipoles. The reverse can also be achieved. Applying a voltage causes polarized molecules to align themselves with the electric field, which, in turn, causes a deformation to develop.

The piezoelectric effect has long been exploited in many different devices. The obviously desirable property wherein a pressure produces a voltage is used in many different ways. In the common doorbell pusher, an applied force produces a voltage, which in turn is used to control an electrical circuit causing the irritating chime or delightful buzz. In the previously mentioned charcoal lighter, application of a force to a piezoelectric device causes an ignition spark. Less obvious to most people, but more widely used, are a whole host of piezo-based devices that serve as small electrically controlled actuators used in a variety of mechanical and industrial situations wherein a small voltage causes a part movement that controls something else, such as a valve.

The piezoelectric effect is literally instantaneous and piezoelectric devices can be quite sensitive to small pressures or voltages. Many microphones based on piezoelectric materials transform an acoustical pressure into a voltage. Alternatively, in piezoelectric speakers, application of an electrical charge causes a mechanical deformation, which can in turn create an acoustical pressure.

Uses can be surprising. Piezoelectric materials have been used in skis to damp out undesirable vibrations that can occur under certain conditions. Here, the piezoelectric effect dampens vibrations by dissipating the electrical energy developed across a shunting. Other situations involving vibratory movements in many products can be selectively damped out using similar technologies.

SHAPE MEMORY ALLOYS

Perhaps surprisingly, eyeglass frames that are amazingly bendable, medical stents for opening arteries that are implanted in a compressed form and then expand to the right size and shape when warmed by the body, tiny actuators that eject disks from laptop computers, small microvalves and a host of other devices, all share a common material technology. The interesting behavior of each of these devices relies upon a phenomenon called the 'shape memory effect' that refers to the ability of a particular kind of alloy material to revert, or remember, a previously memorized or preset shape. The characteristic derives from the phase-transformation characteristics of the material. A solid state phase change – a molecular rearrangement – occurs in the shape memory alloy that is temperature-dependent and reversible. For example, the material can be shaped into one configuration at a high temperature, deformed dramatically while at a low temperature, and then revert back to its original shape upon the application of heat in any form, including by an electrical current. The phenomenon of superelasticity – the ability of a material to undergo enormous elastic or reversible deformations – is also related to the shape memory effect.

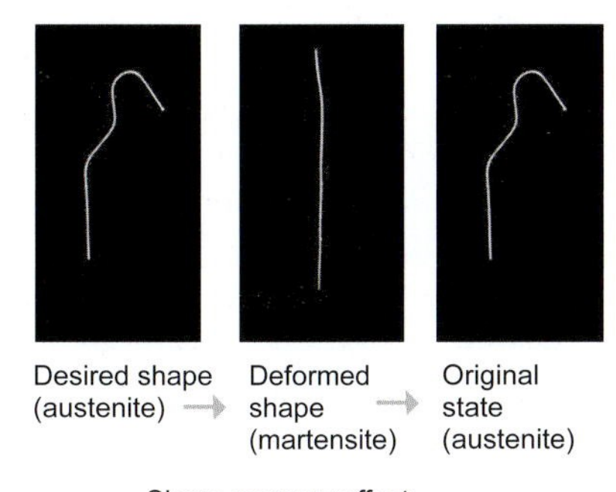

Material is given a shape while in the higher temperature austenite phase

While in the lower temperature martensite phase, the material can be easily deformed into another shape

Upon the application of heat, the material returns to its higher temperature austenite phase and to its original shape

▲ **Figure 4-23** In a thermally induced shape memory effect, a material can be deformed, but 'remembers' its original shape after heating. Shape memory effects may also be induced in other materials by magnetic fields

Nickel–titanium (NiTi) alloys are commonly used in shape memory applications, although many other kinds of alloys also exhibit shape memory effects. These alloys can exist in final product form in two different temperature-dependent crystalline states or phases. The primary and higher temperature phase is called the austenite state. The lower temperature phase is called the martensite state. The physical properties of the material in the austenite and martensite phases are quite different. The material in the austenite state is strong and hard, while it is soft and ductile in the martensite phase. The austenite crystal structure is a simple body-centered cubic structure, while martensite has a more complex rhombic structure.

With respect to its stress–strain curve, the higher temperature austenite behaves similarly to most metals. The stress–strain curve of the lower temperature martensitic structure, however, almost looks like that of an elastomer in that it has 'plateau' stress-deformation characteristics where large deformations can easily occur with little force. In this state, it behaves like pure tin, which can (within limits) be bent back and forth repeatedly without strain hardening that can lead to

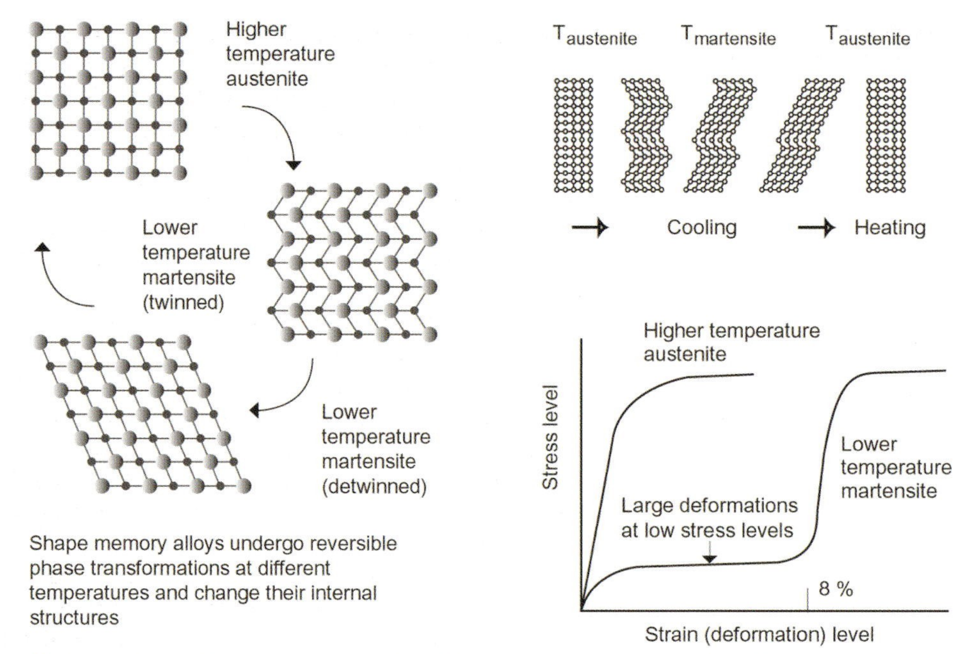

▲ **Figure 4-24** Shape memory alloys (e.g., Nitinol) that exhibit thermally induced shape memory effects

failure. The material in the lower temperature martensite state has a 'twinned' crystalline structure, which involves a mirror symmetry displacement of atoms across a particular plane. Twin boundaries are formed that can be moved easily and without the formation of microdefects such as dislocations. Unlike most metals that undergo deformations by slip or dislocation movement, deformation in a twinned structure occurs by large changes in the orientation of its whole crystalline structure associated with movements of its twin boundaries.

The thermally induced shape memory effect is associated with these different phases. In the primary high temperature environment, the material is in the austenite phase. Upon cooling the material becomes martensitic. No obvious shape change occurs upon cooling, but now the material can be mechanically deformed. It will remain deformed while it is cool. Upon heating, the austenitic structure again appears and the material returns to its initial shape.

A related mechanically induced phenomenon called *super-elasticity* can also take place. The application of a stress to a shape memory alloy being deformed induces a phase transformation from the austenite phase to the martensite

The shape memory alloy changes from an austenite phase to a martensite phase during deformation.

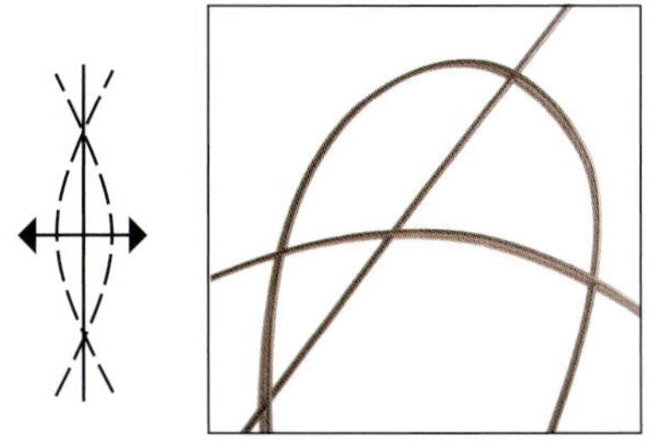

▲ **Figure 4-25** Superelasticity – a mechanically induced shape memory effect

phase (which is highly deformable). The stress causes martensite to form at temperatures higher than previously and there is high ductility associated with the martensite. The associated strains or deformations are reversible when the applied stress level is removed and the material reverts back to austenite. High deformations, on the order of 5–8%, can be achieved. Changes in the external temperature environment are not necessary for the superelasticity phenomenon to occur.

Why these phenomena occur is fundamentally a result of the need for a crystal lattice structure to accommodate to the minimum energy state for a given temperature. There are many different configurations that a crystal lattice structure can assume in the martensite phase, but there is only one possible configuration or orientation in the austenite state, and all martensitic configurations must ultimately revert to that single shape and structure upon heating past a critical phase transition temperature. The process described is repeatable as long as limits associated with the transition phases are maintained. Under high stress or deformation levels, a form of fatigue failure can occur after repeated cycles.

Both of the two primary phenomena associated with shape memory effects – thermally induced effects and mechanically induced effects – have direct applications. In the shape memory effect associated with the thermal environment, a material having an initial shape while in its high temperature austenite phase can subsequently be deformed while in a lower temperature phase martensite phase. When reheated to the high temperature austenite phase by a heat stimulus, such as an electric current (but any heat source will work), the alloy reverts back to its initial shape. During this process, a high force is generated by the phase-changing material. The material can thus be used as an actuator in many different applications. Usually the material provides the primary force or actuating movement as part of a larger device. Since the force and movement occur within the material itself, devices using it are often very simple as compared to more traditional mechanical actuators. Heat in the form of electrical current is easy to apply and electronically control. Hence, the widespread use of shape memory alloys in release latches and a host of other devices.

In the shape memory effect associated with the mechanical environment, or superelasticity, the material can undergo an elastic deformation (caused by an external force) that can be as high as twenty or more times the elastic strain of normal steel. Superelastic materials thus exhibit incredible abilities to deform and still 'spring back' to their original shape. An initial consumer application of superelastic materials was in eyeglass frames that could seemingly be tied in knots, but which reverted to their original shape upon release.

SHAPE MEMORY POLYMERS

Alloys are not the only materials to exhibit shape memory effects. A major effort has been recently directed with considerable success to engineering polymers to have the same effects. Applications are enormous, since polymers can be easily fabricated in a number of different forms. Medical applications, for example, include the development of shape memory polymeric strands to be used in surgical operations as self-tying knots. The strands are used to tie off blood vessels. The strands are given an initial shape, looped around a vessel and, as the body heat operates on the polymer, the strand ties itself into a knot (its remembered shape).

Notes and references

1. Incandescent light is generated by the glowing of a material due to high temperatures, i.e., it is emitted visible radiation associated with a hot body.

2. Flinn, Richard and Trojan, Paul (1986) *Engineering Materials and Their Applications*. Boston, MA: Houghton Mifflin.

5 Elements and control systems

Throughout this book, we have been discussing the unique abilities of smart materials to act locally and discretely in real time. We now know how they function. But how do we decide *when* we want these materials to act and for *what* purpose? Furthermore, *where* do we find the necessary information for guiding the response?

Fundamental to the underlying technological infrastructure of our society are innumerable devices for measuring the extent or quantity of something, or for sensing changes in the state of an object or environment (a simple thermometer pressure gage provides an example here). There are also many devices (transducers) for changing energy from one form to another, e.g., chemical energy into electrical energy. A classic generator converts mechanical energy into electrical energy. Likewise, there are many actuation devices in use wherein energy is transformed into a physical or chemical action. Electrical energy might be used to power a drill or rotate a fan.

Smart materials can assume many forms and serve many of the different roles described. Many of the more basic actions and behaviors of smart materials that were described in Chapter 4 can be directly translated into roles as sensors, transducers, or actuation devices. Indeed, many smart materials of either the property-changing class or the energy-exchanging class inherently provide various sensory functions. It should be evident that the behaviors described are often identical with what more classically defined sensors, transducers and actuators accomplish; but that they do so integrally within the material itself.

A property change in a material that occurs in response to an external stimulus can normally be used directly as a sensor for that same stimulus. We know, for example, that a simple thermochromic material changes its color directly in response to a temperature change. A change in the color of a material, therefore, is a marker of the change in temperature of the surrounding environment. As was previously noted, these same thermochromic materials change colors at specific temperature levels. Thus, colors can be calibrated with temperature levels to provide a temperature measurement device. Since these materials can also be designed to change colors at specific temperature levels, it is quite easy to produce visually evaluated temperature measurement devices. One of the most common examples of this kind of application is the 'thermo-strip' that is placed against the forehead to measure body temperature. A simple visual numerical scale is superimposed on a thermochromic strip, allowing body temperatures to be easily and quickly determined. Other property-changing materials could be similarly used, e.g., a photochromic material as a way of measuring light intensity or a chemochromic material as a sensor for determining the presence of a chemical. Clearly, these kinds of applications are primarily analog devices and do not produce electrical signals that can be subsequently amplified or otherwise conditioned. Hence, their direct use in connection with more complex sensory systems is limited.

The second class of smart materials – energy-exchanging materials – naturally provide both sensor and transducer functions. Some would also provide actuator functions. The classic example here is that of piezoelectric materials. As previously discussed, a force causes a mechanical deformation which in turn causes an electrical energy output to develop, and vice-versa. This material could thus be used, for example, as a force sensor. The output signal from the piezomaterial could be detected, conditioned and interpreted into some useful form. The latter could be a simple numerical display, or the conditioned signal could be run into a logic controller used to govern the complex actions of a mechanical device. Alternatively, an electric current could be used to create a mechanical force directly, with the piezomaterial serving directly as a mechanical force actuation device. Other materials, such as thermoelectrics, could be used similarly. Obviously, thermoelectrics could be used directly as the basis for a thermal sensor and calibrated to be a temperature measurement device. Alternatively, thermal energy could be used as an input to produce an electrical output that could be used to run any other type of electrically operated actuation device.

The goal of assembling smart material components that serve as sensors, transducers or actuators is to form an interconnected whole system that can be activated or controlled to produce an overall intended action or to possess desired response characteristics. Examples that most of us have encountered are those of a thermostatically controlled room heating system or of a sensor-based alarm system. Smart material components may also be used in many other kinds of applications. Several examples are shown in Figures 5–1 to 5–3.

Any complete set of interconnected elements form a system that has particular performance and control character-

▲ **Figure 5-1** This 'electronic nose' contains an array of chemical sensors that swell and shrink, depending on what trace vapors may be present in the air. Measuring this variation allows certain elements in the air to be identified. (NASA)

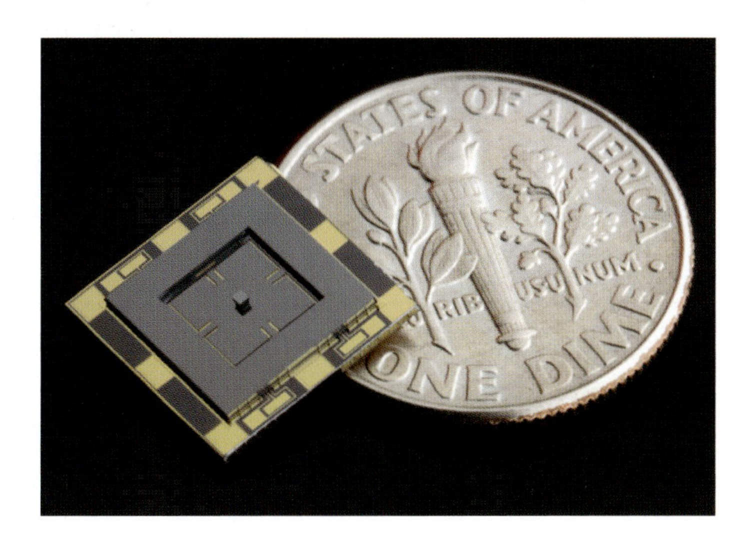

▲ **Figure 5-2** The microgyroscope shown illustrates how MEMS (micro-electromechanical systems) technologies have allowed traditional instruments to be greatly reduced in size. (NASA)

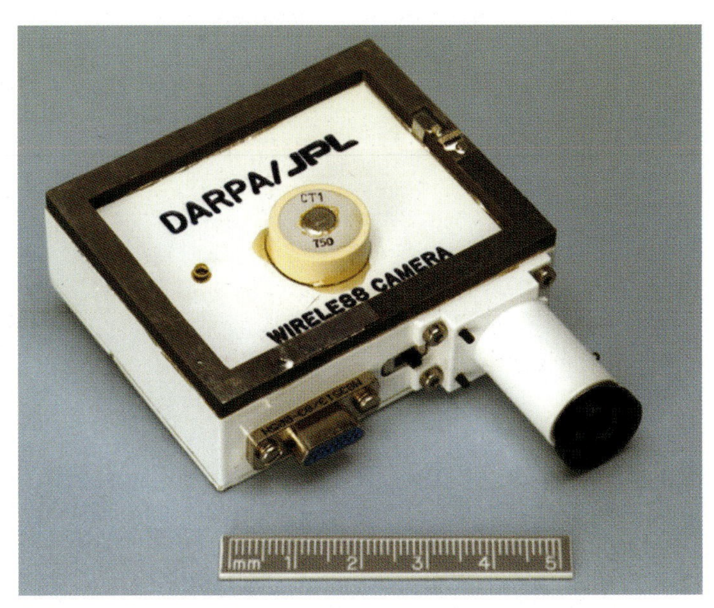

▲ **Figure 5-3** The longest dimension of this experimental wireless camera is less than 10 mm. (NASA)

istics. In terms of a simple input/output model, elements or components in a complete system have historically served different, and often singular, functions. We will see that one of the major attractions of many smart materials is that they can serve multiple functions. Thus, the same material device can be made to serve as either a sensor, an actuator, or sometimes even play both roles.

While we often associate sensor–transducer–actuator systems with process control, and as such they would be of more interest to the engineer than to the designer, there has been some exploration of their possibilities for design. The Aegis Hyposurface by dECOi architects utilizes a straightforward type of position sensor, and then transduces that output signal through a microcontroller to operate a series of pneumatic actuators. Although not seamless, the result is that movement of the body produces a corresponding movement in the wall. This system is representative of the classic mechatronic model that typifies the majority of control applications.

This chapter begins by briefly reviewing basic sensors, detectors, transducers and actuators. Since this field is large, the intent of the review is only to clarify how a traditional system works so that the role of smart materials in this context can be better understood. Overall system control features will also be addressed, including distinctions between closed loop

▲ **Figure 5-4** 'Hyposurface' installation combines position sensors with conventional actuators to create a responsive surface. Images courtesy of Marc Goulthorpe and DeCOI Architects

and open loop systems. The chapter will thus explore the classic mechatronic model that has long dominated system design, but will also look into other models that incorporate smart materials as sensors and actuators. These various models will form the basis for later explorations in Chapters 7 and 8 of concepts such as 'smart assemblies' and 'intelligent environments'.

5.1 Sensors, detectors, transducers and actuators: definitions and characterization

SOME DEFINITIONS

We are used to using certain terms loosely and often interchangeably. Thus, it is useful to begin by more carefully defining certain terms. To *measure* something is to determine the amount or extent of something in relation to a predetermined standard or fixed unit of length, mass, time or temperature, e.g., the length of something is measured in units such as millimeters. There are innumerable *instruments* or *meters* that measure different things.

The term *sensor* derives from the word sense, which means to perceive the presence or properties of things. A sensor is a device that detects or responds to a physical or chemical stimulus (e.g., motion, heat, or chemical concentration). A sensor directly interacts with the stimulus field. In contrast with a measurement device, a sensor invariably involves an exchange of energy or a conversion of energy from one form to another. In normal usage, the term sensor also signifies that there is an output signal or impulse produced by the device that can subsequently be interpreted or used as a basis for measurement or control. A typical measurement device such as a meter-stick, however, is not a sensor.

Sensors and transducers are closely related to one another since both involve energy exchange. A *transducer* is normally a device that converts energy from one form to another, e.g., mechanical energy into electrical energy, although a transducer can also transfer energy in the same form. Transducers are normally used for the purpose of transmitting, monitoring or controlling energy. By contrast, sensors – which also involve energy exchange – interact directly with and respond to the surrounding stimulus field.

As usually used, the term *detector* refers to an assembly consisting of a sensor and the needed electronics that convert the basic signal from the sensor into a usable or understandable form. An *instrument* is a device for measuring, recording or controlling something. An *actuator* is a device that converts input energy in the form of a signal into a mechanical or chemical action. This term typically refers a device that moves or controls something; most frequently, an actuator produces a mechanical action or movement in response to an input voltage.

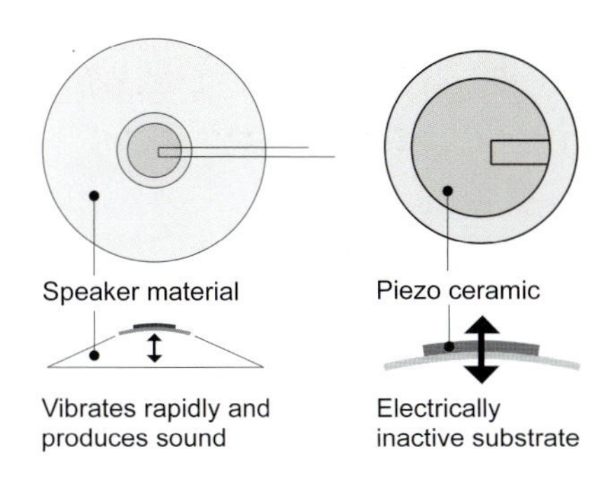

▲ **Figure 5-5** A common small piezoelectric speaker. It is based on the actuation capabilities of piezoelectric materials

In an actuator, an external stimulus in the form of an input signal (such as a voltage) produces an action of one type or another. In a sensor, an external stimulus (such as a mechanical deformation) produces an output signal, often in the form of a voltage. The signal, in turn, can be used to control many other system elements or behaviors. We will see that in many cases the same device that serves as a sensor can also be reconfigured to serve as an actuator. This is certainly the case, for example, with the piezoelectric devices discussed earlier. A speaker based on piezoelectric technologies is shown in Figure 5–5.

Measurement

Measurement is the determination of the amount or extent of something in relation to one of four standards – length, mass, time or temperature. These standards are predefined, e.g., the definition of a meter is defined as the distance traveled in a vacuum by light in 1/299 792 458 seconds, and are carefully maintained by various standards agencies. Other standards are derived from these basic four. Measurements based on these standards can be made independently of the nature of the surrounding environment. Unlike sensor outputs, measurements are not relative to the surrounding environment or stimulus field. Measurement instruments or devices either provide a way of directly comparing something to a standard (e.g., a ruler), converting something to a standard (e.g., a manometer) or converting a measured quantity to an interpretable signal.

Common measurements are related to mechanical, thermal, electrical, magnetic or radiant energy states. Length, area, volume, time and time-related measures (velocity, acceleration), mass flow, torque and others are related to a mechanical environment. Temperature, heat flow, and specific heat are related to a thermal environment. Voltage, current, resistance, polarization and others are related to an electrical energy state. Field intensity, flux density permeability and others are related to a magnetic environment. Phases, reflectances, transmittances and others are related to the radiant energy environment. Concentration, reactivity and similar measures are related to the chemical environment.

SENSOR TYPES

There are many different types of sensors and transducers. A basic way of thinking about the different types is via the energy form that is initially used – mechanical, thermal, electrical, magnetic, radiant or chemical. Sensors and trans-

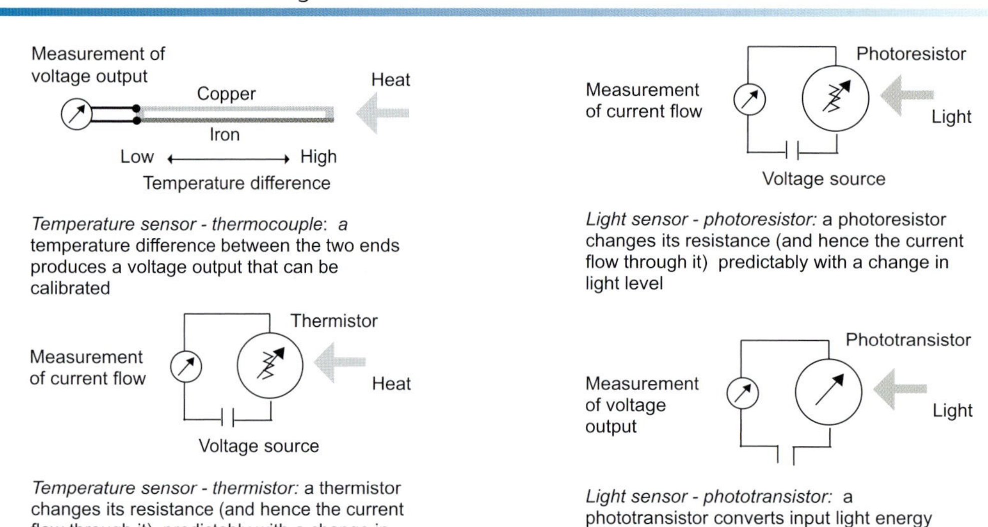

▲ **Figure 5-6** A sampling of different temperature and light sensors. Small signal amplifier circuits are also often incorporated

ducers can be based on any of these energy states. Another way of thinking about the different types that exist is based on their expected usage, e.g., proximity sensors or sound sensors. Here we look briefly at several of the basic types that are particularly relevant to design applications.

Light sensors

Numerous types of light sensors exist. Semiconductor materials provide basic technologies. Radiant energy in the form of light striking a semiconductor material produces a detectable electrical current *Photodiode sensors,* for example, can be connected to a microprocessor to provide a digital output. Light levels can thus be not only monitored but logged as well. *Phototransistors* that convert radiant energy into voltage outputs are used for light sensors as well and form a kind of switch based on the amount of incident light. Other forms are based on the use of various kinds of *photoresistive* materials. Often called *light-dependent resistors* (LDRs), these resistors change their value according to the amount of light falling on them.

Infrared sensors are based on a form of phototransistor that normally involves both an infrared source (such as in infrared LED or infrared laser) and an infrared receiver (such as a photodiode or photoresistor). A reflectance light sensor typically has both an infrared light-emitting diode (LED) and a photodiode. The LED emits non-visible infrared light and the photodiode measures the amount of reflected light. (See also *Semiconductors, and Photovoltaics*).

Sound sensors

The most common type of sound sensor is based on the use of piezoelectric materials. In a piezoelectric material, a mechanical force produces a measurable electrical current. In a sound sensor, acoustical sound wave pressures produce a force in a piezoelectric material in a microphone, and a detectable current is thus generated.

Thermal sensors

Specific technologies for detecting changes in the thermal environment include various classic thermometers, thermocouples, thermistors and others. Several are mechanically based. A room thermostat, for example, works by the exchange of thermal energy to induce the bending of a bimetallic strip, in which the two metals have different thermal expansion coefficients. The bending, in turn, generates some sort of output signal. How this is done varies, but can range from triggering a simple switch to generating an electrical signal via strain gages (see below) or other electrically based devices. Thermistors, by contrast, are resistors that change their electrical resistances in a predictable way with a change in environmental temperature. Any change in resistance can in turn be detected by any of a variety of electrical circuits, which in turn can be converted into a digital display output.

Humidity sensors

Measurement of absolute and relative humidity levels is a common environmental need. Measurements can be difficult, however, because of the way air pressure, temperature and moisture content in the air interrelate. The classic psychrometer evaluates relative humidity by measuring the temperature difference between a 'wet bulb' and a 'dry bulb' thermometer. Other technologies include different capacitive or impedance devices. In impedance devices, the resistivity of a moisture-absorbent material changes with the amount of moisture present. Its impedance or resistivity is then measured. A capacitance device is used that has a moisture absorbent material whose dielectric properties change with the absorption of moisture. The resulting change in the dielectric properties corresponds to the capacitance of the material, which in turn can be easily measured. In an electrolytic device, the moisture present can be measured by the current need to electrolyze it from a desiccant. Other kinds of technologies can be used as well.

Touch sensors

The role of touch sensors is intuitively obvious. There are, however, many different types of touch sensors. Many are based on simple mechanical operations of one type of another. Simple touch switches, for example, that open or close electrical contacts are a form of touch sensor. Others involve more sophisticated forms of touch measurement. A capacitance-based device, for example, is based on the phenomenon that when two electrically conductive materials are near one another, an interaction occurs between their electric fields that is, in turn, measurable.

Touch pads commonly found on many devices can be based on several different technologies, including capacitive ones. In a capacitive approach a surface is made of a small grid of conductive metal wires, which are in turn covered by an insulating sheet or film. A finger touching the pad creates higher electrical capacitances in the crossed wires directly beneath the touch. These increases can be detected and used to determine the finger location.

In design applications, *membrane switches* that include touch pad technologies are extensively used for commonly found applications such as in microwaves and other products. These devices can be relatively simple flexible switches or contain extensive circuitry for performing complex functions. Briefly, membrane switches typically consist of a laminated assembly with a switch layer and a thin circuit board separated by a spacer, adhesive carrier films and covering face plates. Touching the exterior surface causes an electrical connection to be made between the separated switch layer and circuit layer. Additional layers can be incorporated for graphics and backlighting. Capacitive technologies are often used, but even simple pressure switches can be employed as well. LEDs can be built into one of the circuits. These switches can be rigid or flexible and are easily produced. They are not as rugged as other types of switches.

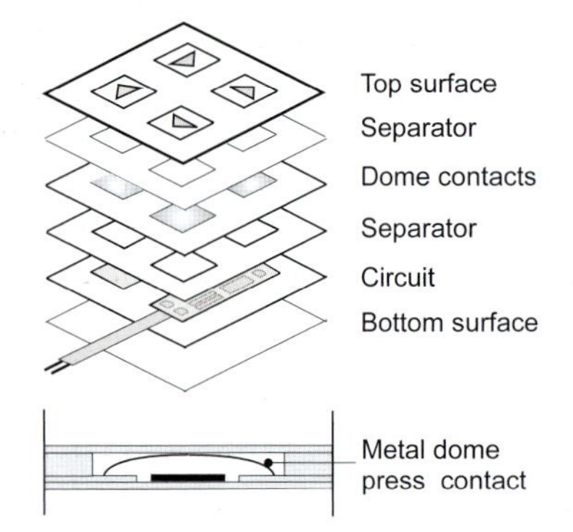

▲ **Figure 5-7** A simple membrane switch that is found on many common consumer goods

Several different approaches are in use for touch screens. In the resistive touch screen, switch areas are formed on the screen, typically in matrix form. A typical device, for example, might consist of two laminated sheets, one with conductive strips in the *x* direction and the other with conductive strips in the *y* direction. Lines can be made in many ways, including through the etching of transparent tin oxide (ITO) coated

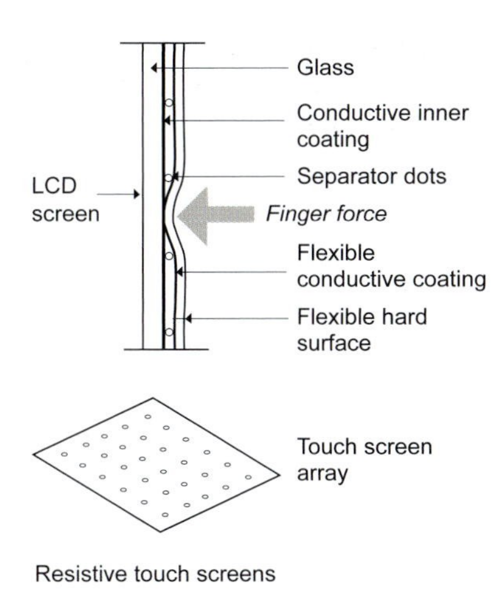

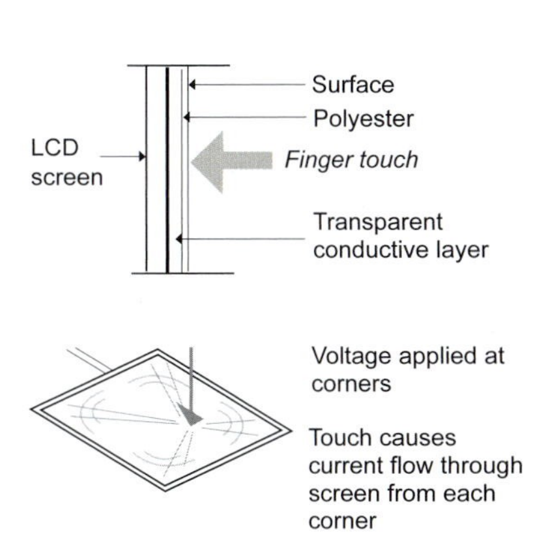

Capacitive touch screen - measurements of current flows allow position to be determined

▲ **Figure 5-8** Touch screens: two common technologies for touch screens are shown. Other technologies are possible as well

sheets. Touching a point on the screen causes a contact to be made that is then detected by a circuit. These types of screens are simple and rugged, but resolution is limited. Other approaches use dual panels with an applied voltage. Touching one panel causes a voltage drop, which is then detected by the other panel and this information is used to determine the touch location. These screens can be high-resolution. Touch screens can be based on capacitive technologies as well. In a typical application, electrodes are distributed across a screen's surface and a voltage is applied to screen corners. A finger touching the screen creates a detectable condition. A controller can calculate the finger's position from current measurements. Other touch screen approaches can be based on surface acoustic wave technologies.

Position sensors

These sensors are used to determine the location or placement of an object, or its direction and speed of movement. A wide array of different types of position sensors are available that are based on different mechanical, optical, or inductive/capacitance technologies. There are position or velocity sensors, force or pressure sensors, mechanical deformation sensors (including strain gages), and vibration or acceleration sensors. Many of these sensors are found in industrial operations.

Proximity sensors

Proximity sensors can be used in many different ways; to indicate whether a physical element is near another one – there are a vast number of applications in industrial machines here – or in quite different applications such as some simple door-closing devices. Normally, the term 'proximity sensor' is used in situations where the distances involved are relatively small, e.g., in assembly machines, and less so for larger-scale applications such as 'obstacle detection' (see below). They also typically sense the presence of an object, not its range.

Mechanical proximity sensors are often used when physical contact between parts is both possible but not problematic. They are very easy to use and reliable. Optical proximity sensors are non-contact technologies based on the use of a light source (such as an LED) and a light detector (such as a photosensitive device such as a photoresistor). Some project light, others use fiber-optic cables to deliver light. These devices are useful when there cannot be contact between adjacent elements. They respond quickly, but their perfor-

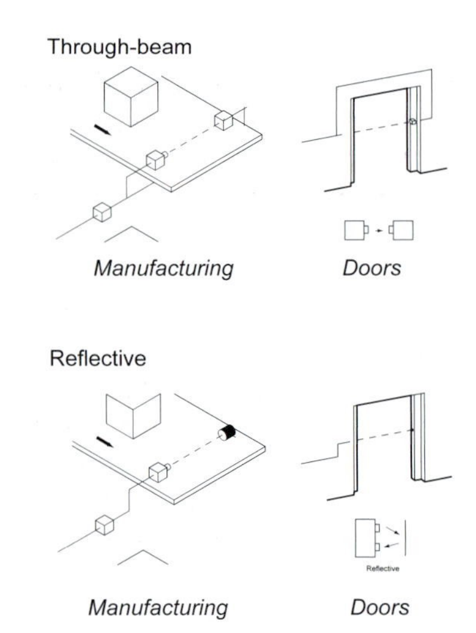

▲ **Figure 5-9** Applications of different sensors in manufacturing and other contexts (detection of presence, counting, etc.)

mance can be hampered when there is dust or moisture in their operating environment. There are other light-based proximity sensors as well and many different kinds of applications. Of particular interest is the 'light curtain'. Here an array of parallel infrared beams is projected across a space to a receiving unit. When the light is interrupted by any solid object, including a human, the light curtain controller sends out a signal activating some other operation. These devices are often used around hazardous equipment.

Ultrasonic position sensors find use in large-scale applications. They are often used as 'obstacle detection' systems. They can be small and relatively inexpensive. They find use in everything from robotic devices to consumer devices such as digital cameras. A typical ultrasonic ranging system sends out a short pulse of high frequency sound waves in a cone towards the object. The sound reflects back causing a responding voltage output in a transducer that is a direct analog to the distance to the object. The cone of vision can be varied, but is an important design factor. Ultrasonic sensors of this general type can normally function in environments where there is heavy dust and moisture. Thus, they have been used in automated 'car washes'. Surrounding high noise levels can sometimes be problematic. Inductive and capacitance sensors, by contrast, are used in short range situations, but can be very reliable and also detect the presence of something without physical contact.

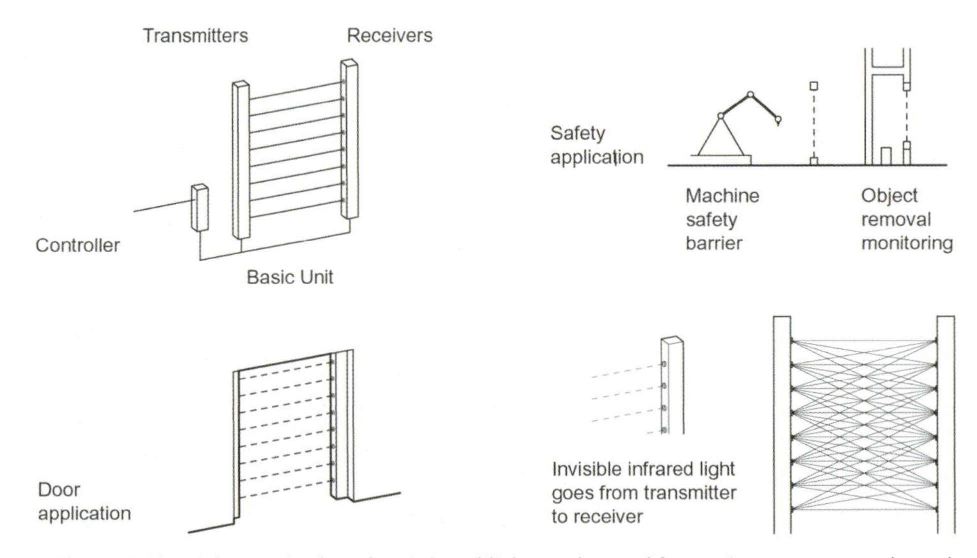

▲ **Figure 5-10** Light curtains based on infrared light can be used for sensing movements through openings. Machines can be cut off, for example, if something interrupts the infrared light curtain

Motion sensors

Motion sensors are obviously used in a wide array of different settings, including in common home security systems. Most of these types are based on the use of infrared technologies, and thus primarily detect the heat differential between a moving object (e.g., a person, an animal, or an engine) and the surrounding environment. Filters for these sensors can be devised so that the sensor responds only to the heat or infrared radiation generated by humans, which occurs in a fairly narrow wavelength band. In a typical infrared sensor system, there are dual projected infrared cones that are more or less parallel. Comparable ambient conditions are sensed within each cone and the unit logic system essentially cancels them out. A human in motion, however, that passes through

▲ **Figure 5-11** This microseismometer is capable of sensing accelerations. Thus, it can be used for sensing vibrations. (NASA)

either one of the cones – it is not possible to pass through both simultaneously – creates an unbalanced condition that results in an output signal that in turn sets off an alarm or generates some other action. Other technologies in use include those based on the use of photoresistive devices of one type or another. Again, typically more than one device is used so that signals are generated by moving objects when they upset a state of balance during ambient conditions.

Other types of motion sensors include various accelerometers, which include vibration-sensing devices. These work in a variety of ways, including via the use of piezoelectric materials. A small mass would be connected to a thin piezoelectric plate. Its inertial movements associated with accelerations or vibrations would be transmitted into the piezoelectric plate. The deformations induced would in turn cause voltage outputs that can be detected and calibrated.

Chemical, magnetic and other basic sensors

A whole series of sensors have been developed to detect the presence and/or concentration of one or another type of chemical. They invariably involve a reaction of one type or another between a compound embedded on the sensor and the actual chemical environment being monitored. Outputs vary. Common litmus paper, for example, is one form of a simple chemically based sensor in which color change is the output. In other more sophisticated sensors, outputs are converted into measurable electrical signals. Likewise, there are a whole host of sensors that in one way or another respond to the presence and/or intensity of a magnetic field. These almost invariably have electrical signal outputs.

Many of these sensors are used in either industrial process applications or in environmental sensing. For detection of air pollutants, for example, there are many kinds of instruments that measure specific concentrations of pollutants, including both gaseous and particulate matter. Gas monitors can have sensors that respond to hydrocarbons, ammonia, sulfur dioxide and a wide variety of other gases. Some systems use interchangeable sensors. Special purpose monitors have been developed to quickly measure concentrations of specific gases such as carbon dioxide. Since various interactions can be important, some gas monitors are packaged with temperature and humidity sensors or others as needed. A similar range of instruments is available for pollutants found in water. These are often in turn used with flow meters and other related instruments.

Environmental sensors

The term 'environmental sensors' does not describe a unique group of specially devised sensors. Rather, the term is used to describe a loosely defined group of sensors – including many already described above – for testing or measuring changes in the broad environmental conditions surrounding habitations, whether at the room, building or urban level. The importance of environmental considerations in any built project needs no explanation in this day and age. A myriad of corresponding regulations have also arisen that reflect these concerns.

Primary to any environmental assessment is a clear understanding of existing environmental conditions and the way they change over time. A host of sensor-based devices have been developed to both initially assess and continuously monitor these conditions. For the air environment, various assessments needed include temperature, humidity and other basic measurements such as volumes, rates and directions of air movement. Other sensors are needed to detect different concentrations of gaseous and particulate matter. Thus, many different types of measures are recorded by 'environmental sensors.' Most come equipped with various data-logging devices for use in field circumstances. What 'environmental sensors' have in common is that they are of collective interest as tools for monitoring changes in our environment rather than that they have any common technological basis.

Biosensors

In its broadest definition, a biosensor is any sensing device that either contains or responds to a biological element. Generally, though, because the field has become so large, most modern definitions distinguish between the two types, and the term biosensor is now more appropriately applied to a sensor that *contains* a biological element. Although we currently think of biosensors as highly sophisticated micro-electronic devices, biosensing can probably be traced back to an infamous precedent – the canary in a cold mine. Supposedly, coal miners brought canaries with them into coal mines as the tiny bird's metabolism was such that even a trace of carbon monoxide in the atmosphere would be enough to cause the poor bird's demise. Today's biosensors are premised on this same highly specific selectivity, although animal cruelty is fortunately no longer required.

The biological element is the primary sensing element; and in the most common types of biosensors this element responds with a property change to an input chemical.

Originally developed for sensing blood glucose levels for diabetics, biosensors have expanded into applications as diverse as process control and food inspection. Regardless of the application, the focus of the biosensor is molecular recognition of a chemical, also known as an analyte.

Biological systems are highly adept at molecular recognition; for example, enzymes in the human body are capable of responding selectively and specifically to individual components in our surrounding environment. Indeed, enzymes are the most commonly used biosensing element, but antibodies, cells, microbes and living tissues can be incorporated as well depending on the analyte. The key requirement in choosing the biological element has to do with its ability to provide a selective response through binding to the analyte at the expected concentrations, regardless of the other chemicals that may be present or of an inhospitable environment. When binding occurs, the biosensing element may respond in several ways, from conversion to another chemical, or release of a chemical, but the most useful manner is if the response results in a change of one of its electrical or optical properties. The transducer element is responsible for converting the element's response into a measurable signal. Many of today's biosensors utilize semiconductors as transducers; the biological element is deposited on the semiconductor surface and thus electron flow is directly affected when binding to an analyte occurs.

Biosensors may be representative of the concept of a smart material at its best. Chemical analysis had been one of the most cumbersome and time-consuming of all measurement processes. Cultures had to be grown in a laboratory, or samples were collected and processed through a gas chromatograph to split out the important components that were then characterized by mass spectrometry. Biosensors eliminate the need to collect and process samples, as they do it in situ, and the time element, which previously took days, is collapsed into real time. A large, slow, indirect process has been replaced by a small, fast, direct process. As a result, biosensors are currently the fastest growing segment of the sensor industry.

Swarms (smart dust)

Biosensors gave rise to the concept of swarms or smart dust. The ability of a biosensor to respond to even a single molecule of a particular chemical renders the technology ideal for environmental monitoring, particularly in unknown or hostile environments. The issue, however, that plagues all sensors, and particularly biosensors, is the location of the sensor relative to the measurand. Objects that are relatively confined or environments that are highly predictable pose little problem, but unknown environments in which the chemical of concern may be located in pockets or moving in an unpredictable pattern are unlikely to produce usable information. Just because a sensor is capable of responding to a single

▲ **Figure 5-12** Smart dust. The visions for smart dust presume that it will be relatively undetectable by the human eye. Current research efforts have dramatically brought down the size of the dust. (Berkeley)

molecule does not mean that it will encounter that molecule. Many early proposals for the deployment of environmental monitors called for establishing a distributed network of multiple sensors to overcome this problem, but this too had its limitations. Sensors were tethered to their locations, requiring a large electrical infrastructure, and thus the system could not adapt to shifting conditions.

The concept of 'controllable granular matter', or, more popularly, 'smart dust', is a provocative one: tiny sensors could be dropped from planes, spreading over an area much as dandelion seeds do with one quick breath. Obviously, there are many physical obstacles: the need for further miniaturization of the sensor, their aerodynamics, their distribution coverage and how they communicate. The initial physical challenges of 'smart dust' are beginning to be overcome, but it is the informational challenges that are currently looming large for researchers. A parallel research effort in wireless networks is addressing not only the communication between the sensors and the data collection site, but also intra-sensor communication.

Object tracking and identification systems

While not really sensors in the normal meaning of the word, there are several technologies that serve sensor-like functions which are used for determining the presence of objects, as well as identifying and tracking them. Radio Frequency Identification (RFID) tags have become widespread and almost ubiquitous. These tags listen for a radio query and respond by transmitting an identifiable code. A typical RFID system consists of a central transceiver connected to a computer-based processing device and a transponder, or tag on the object that is being tracked. The transponder contains a small integrated circuit that picks up radio signals and responds with identification data. Typically, the latter need not contain batteries. High frequency RFID systems can transmit fairly long distances. Problems include common radio technology problems such as interference, absorption (e.g., by water or human bodies at certain wavelengths), and available bandwidth.

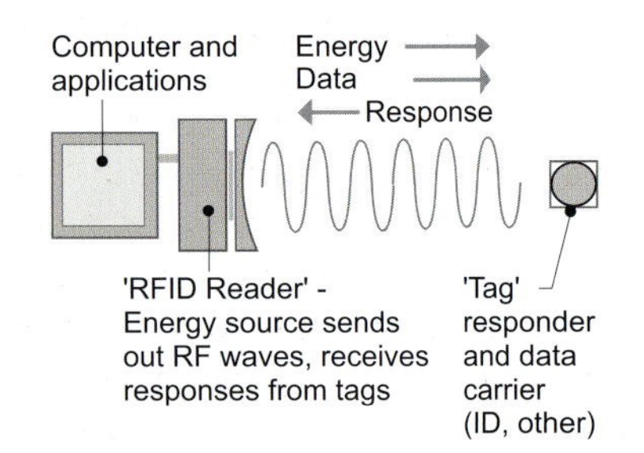

▲ **Figure 5-13** RFID (Radio Frequency ID) tags and reader

RFID tags are inexpensive and can be placed virtually anywhere. Hence, they find wide application in everything from inventory control applications, counting and charging (e.g., for automobile tolls), process applications (e.g., stages in manufacturing) and so forth. Since RFID tags depend on radio technologies, they obviate the need of other tracking systems, such as the all-familiar bar coding, for line-of-sight readings.

5.2 Control systems

BASIC FEATURES

We have so far briefly explored how different components of a complex system work. Here we look at how more complex assemblies function. This is a huge topic that is ultimately the domain of the engineering profession and beyond the scope of this book. Here we only look at basic characteristics and issues critical to how designers use these systems.

At the risk of excessive oversimplification, a typical system of sensors and actuators that are intended to accomplish specific tasks consists of elements providing a number of different functions:

- Sensors and transducers
- Signal conditioners
- Transmitters/converters/receivers
- Logic controllers
- Displays/recorders/actuators

In the traditional *mechatronic (mechanical-electronic)* model, several of these functions are provided by individual components that are interconnected and provide an overall desired response. The roles and operations of sensors, transducers and actuators have already been briefly explored. The output of a sensor that is responding to some energy stimulus may or may not be in a readily usable or interpretable state. Transducers may or may not be needed to change the sensor output signal to some other energy state. The output signal is also usually in need of further conditioning to boost its amplitude, filter out unwanted 'noise' or other reasons. The conditioned signal invariably needs to be transmitted elsewhere so that it can be used directly as an input to something else, which in turn means that it has to be received elsewhere. Transmitting and receiving devices are thus obviously needed. These processes may again require signal conversion. Actuation devices could then be directly activated.

In the *enhanced mechatronic* model that is more sophisticated, transmitted signals that have been conditioned are first manipulated according to a logical intent and then transmitted to final actuators. Of special interest here are the logic controllers. It is here that the system is ultimately given its directions. A designer might want a motion detector to cause an alarm to go off when movement is detected, or a door to open when motion is sensed. Operations of this type can be done using actual hard-wired circuits that use common

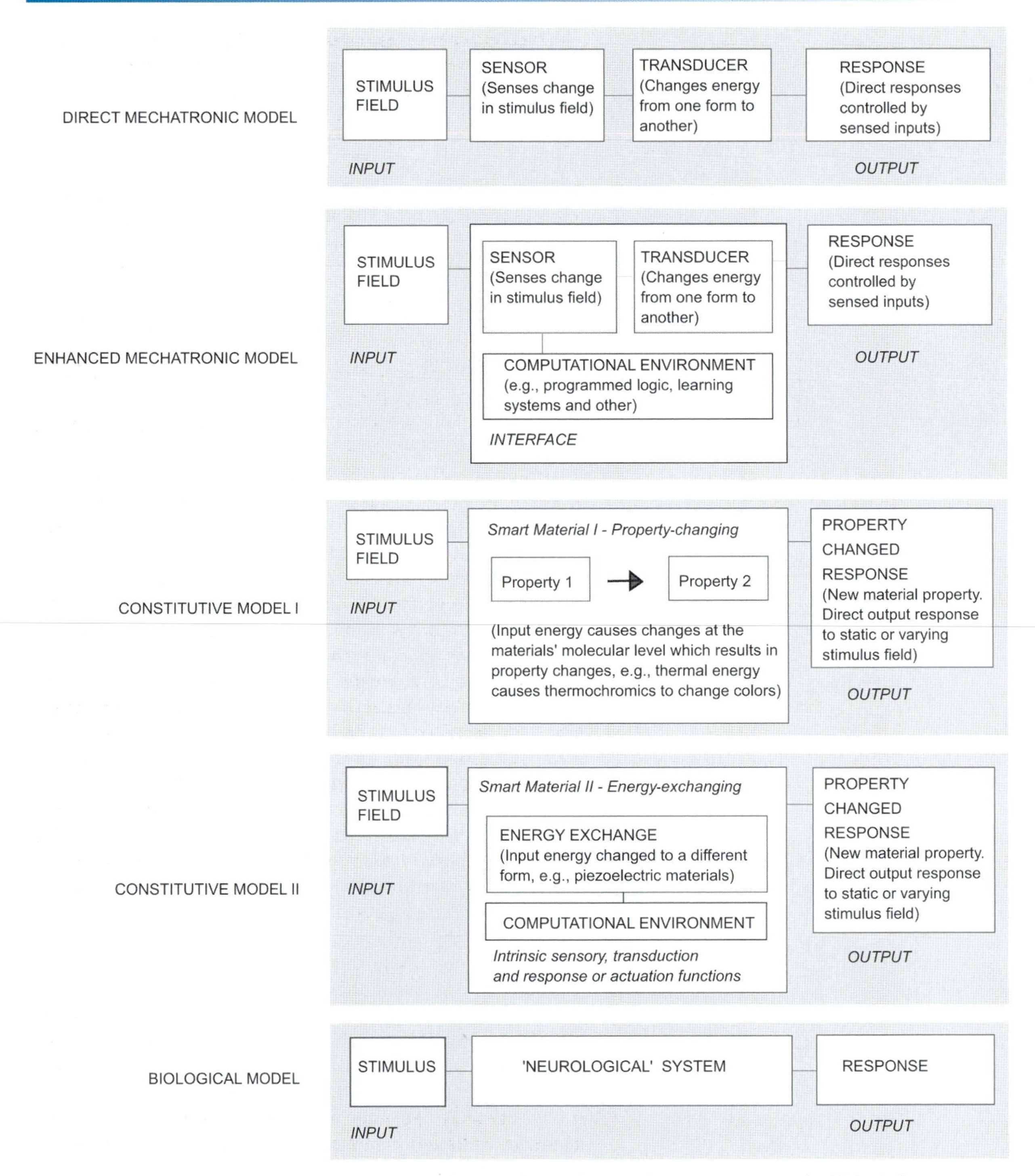

▲ **Figure 5-14** Differerent input/output control models (from the simple sensor/actuator system to the biological system with its intricacies)

electrical devices to perform a series of operations, such as an inexpensive timer circuit where a charge builds up in a capacitor and is then periodically released, which in turn causes some action to occur. Circuits with surprisingly intricate logic capabilities can be built in this way.

An advantage of the use of both property-changing and energy-exchanging smart materials within the context of sensor/actuator systems is that many of the actions described occur internally within the materials themselves. In some cases the sensor/actuation cycle is completely internalized. In other cases, additional elements may be required for certain kinds of responses, but the complexity of the system is invariably reduced.

For Type 1 property-changing materials, there are intrinsic and reversible property change responses. The *constitutive model I*, shown in Figure 5–14, can be used to describe basic input and output relations for Type 1 smart materials.

When the full advantages of a complete control system are desired, including programmable logic capabilities or closed loop behaviors, it is clear that energy-exchanging smart materials that can generate electrical signals would be a seamless improvement. While the behaviors of energy-exchanging materials are not normally programmable in the accepted sense of the word, they can easily become part of a complex system and still serve to reduce its overall complexity. The constitutive model II, shown in Figure 5–14, can be used to describe basic input and output relations when these kinds of Type 2 energy-exchanging materials are used.

It is interesting to note that more and more research is directed towards making as many overall system behaviors as internalized as possible. Ultimately, smart materials offer the possibility of making the overall system seamless. The biological model diagrammed in Figure 5–14 suggests the fundamental nature of this kind of completely internalized environment.

OPEN AND CLOSED LOOPS

For complex systems or where there is a need to frequently change logical parameters, logic controllers inherently involve either direct connection to an external computational environment or do so via an internal microprocessor. For example, a designer might want an activated motion detector to cause a series of different lights to blink at different rates. A simple program could be written that takes the conditioned signal as an input and then uses it as a trigger to control different output signals to the final lights. The sequence and rate of light activation could be directly programmed. Many of these programming environments use common interpreted programming languages, e.g., Basic or C, while others use more direct programming techniques. Once a programming environment is introduced, the control possibilities are exciting since all manner of operations can be envisioned (see *Microprocessors* below).

A control system can have many features. Of particular importance herein is the distinction between open loop and closed loop systems. In a simple open loop system, the input is processed to produce a desired response. Thus, a sensor might cause a slider on a mechanical device to move 25 mm or for a shaft to twist 15°. In another case, a sensor might cause an actuator to open a door. In an *open loop* system the input system may be processed and an output signal sent to an actuator to cause the desired action, but there is nothing inherent in the system that checks to see if the desired action actually took place. Did the slider move 25 mm or did the door actually open? If something prevented the final action from occurring, there is no way that the system would know this.

A *closed loop* system has additional features that allow the system to check to see if the intended output action did indeed occur in response to the input as desired and anticipated. Thus, in the door opener or slider example, there might be some type of position sensor or limit switch that detected whether the door was in the open position or that the slider moved 25 mm as desired. This information would be fed back into the system and the original state compared to the final state. The comparison would then indicate whether the desired action took place. If not, the system could be programmed to 'try again,' issue an error message, or do something else. Feedback systems of this type normally involve use of microprocessors because of the logic involved, but a review of the history of automation suggests that there are many other ways, including mechanical means, by which feedback control can be obtained.

MICROCONTROLLERS

The primary purpose of a *microcontroller* is to communicate with an electronic device and control its actions. Microcontrollers are based on microprocessor technologies. Microcontrollers are typically unseen to users, but they are buried in innumerable devices found in diverse settings such as automobiles, home appliances or video equipment. A microcontroller can be programmed to execute instructions in a desired way, and thus to follow a series of sequenced operations that control the actions of a linked device. Microcontrollers can be designed to be stand-alone devices that execute pre-programmed actions from memory, or they can be designed to interface directly between a primary user-controlled computer and an external device.

A microcontroller converts inputs into outputs. A microcontroller typically receives input signals and produces output signals that control an electrically activated mechanical device. Microcontrollers invariably involve both hardware components and software that interprets inputs and controls outputs. Microcontrollers come in a variety of types, sizes and capabilities. A typical microcontroller might consist of a small *microprocessor* that has computing capabilities, a built-in capability for storing in memory a programming language interpreter (e.g., for Basic or other programming languages), a series of input/output (I/O) pins to link to both input and output devices, and a built-in power supply or connection to an external source. A typical microcontroller can be reprogrammed at will via connection to a primary computer that houses the primary programming language. Typically, programmed instructions relate to how information from the input pins (say an electrical signal from a sensor) is manipulated and what signals are sent to each of the output pins, which in turn control connected devices.

5.3 MEMS (micro-electro-mechanical systems)

This idea of seamlessness has propelled the development of integrated sensor–transducer products toward the incorporation of computational intelligence. Smart materials and microtechnology had adhered to parallel, albeit close, research and development tracks. In a curious crossover from the information world to the physical world, a ubiquitous material emerged as the means to merge the two worlds. Silicon, the workhorse material of semiconductors, revolutionized the communications and electronics industries when it was introduced due to its rather spectacular electrical properties. But silicon may well be even more compelling as a mechanical material. Three times stronger than steel, but with a density lighter than that of aluminum, silicon also has the near ideal combination of high thermal conductivity with low thermal expansion. On a dimensionless performance level, silicon outperforms all other traditional mechanical materials. Unique about silicon, however, was that there was an entire

▲ **Figure 5-15** Two views of a spider mite crawling across the surface of MEMs devices. The top view is of a comb drive, the bottom view is of a gear chain. (Images Courtesy Sandia National Laboratories, SUMMiT TM Technologies, www.mems.sandia.gov)

fabrication industry already tooled for the manufacturing and machining of silicon components at the micro-scale. Remarkable structures could be directly machined on a silicon chip, including microscopic gears, levers, drive trains and even steam engines! Millions of elements could be combined in a device no larger than a postage stamp. Thus was born a micro-machine that had simultaneous electrical and mechanical functions.

The term micro-electro-mechanical systems (MEMS) has come to describe any tiny machine, but the more precise definition is that a MEMS is a device that combines sensing, actuating and computing. The earliest MEMS tended to lean toward one or another aspect, rather than equally addressing all three, such as the *sensing primary* accelerometers for air bag deployment, the *actuating primary* ink jet printers, and the *computing primary* analyzers for chemical analysis. Many of these early applications did not truly exploit the true potential of MEMS as a smart system, rather the fabrication capabilities simply allowed for miniaturization of more conventional equipment. Today's MEMS have much higher expectations, as they are being developed for unprecedented capabilities: navigation and control of unmanned flight, remote evaluation of the changing characteristics of environmental hazards such as forest fires, and implanted analysis and control of biological processes.

One of the most interesting directions is the development of micro energy systems. A common problem among all electronics, systems, machines and any material with an electrical need is the provision of power. Regardless of how small, how direct and how distributed a component may be, electricity must still be supplied. When any device is miniaturized, its power needs, in terms of both voltage and current, are greatly reduced, but our traditional power supplies can not be correspondingly reduced in size. Batteries have become a fairly standard accompaniment, often dwarfing the component being powered. Micro-machines can perform almost any task on a smaller scale than a full size machine can do on a larger scale as long as the rules for kinematic and dynamic scaling are adhered to. Smart dust could be part of a MEMS device with a rotor, and thus be able to fly to desired locations. And unlike many tasks that require large amounts of force or power and cannot be scaled down, such as an automobile drive train, the electron movement inherent in electricity has no such large-scale needs. A MEMS device may need just a few milliwatts of power, which can be easily achieved with tiny generators. Labels associated with building-size HVAC equipment are now routinely associated with MEMS energy devices – we now have micro-compressors, chillers, heat pumps, turbines, fuel cells and engines. While much of the early impetus for micro energy systems was for the replacement of batteries, there is growing interest in exploiting the energy transfer capabilities

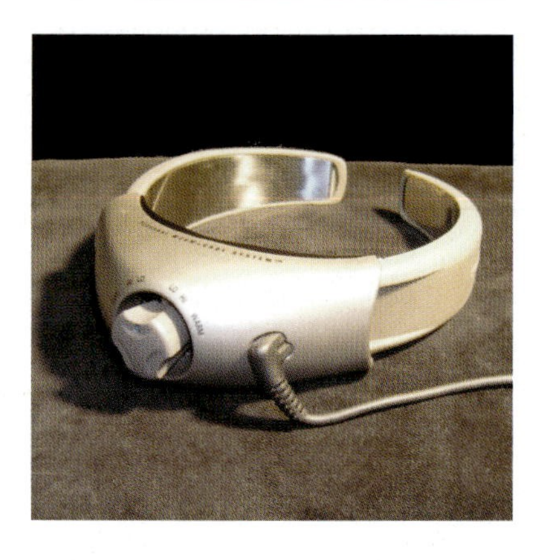

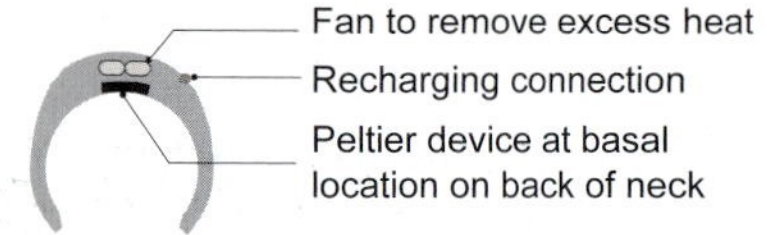

▲ **Figure 5-16** This 'personal cooling and heating' device based on Peltier technologies is worn around the neck

of these devices directly. For example, one of the defined goals for micro-power was to replace the heavy battery needed for the portable, albeit unwieldy, cooling systems that soldiers wear in extreme heat conditions. If the power supply could be miniaturized, then why not the cooling system itself? Scaling of thermal behavior is much more difficult than that of kinematic behavior, nevertheless there are large research efforts currently proceeding in this area. (These energy systems will be discussed in greater detail in Chapter 7.)

5.4 Sensor networks

If remote or local power generation will allow systems to become more autonomous, networks and webs are intended to make them more interactive. Smart dust enables the wide distribution of sensors and devices, but there needs to be a corollary effort in how the information they gather is processed and then acted on. Obviously, a single particle of smart dust could combine all these activities into a single MEMS device known as a 'mote' with each one communicating to a central station through wireless technology. But if we were to push the idea of 'smartness' to its fullest potential, then the motes would communicate with each other and collaboratively decide which one needs to take action. This is precisely the tenet behind the smart sensor web that many agencies are developing, including NASA and the Department of Defense.

For smart sensors to be effective in monitoring environmental or battlefield conditions, tens of thousands of them would need to be distributed. Addressing each sensor individually would flood data banks with generally useless information that must be sifted through. NASA tackled this problem in their design of the Mars Rover, when they recognized that fully networked communication between all the components, far from being usefully redundant, actually increased both the opportunity for failure and the severity of the consequences. If, however, clusters of sensors communicated among each other, decisions could be made locally and directly. There are many models for this, from treating sensors as individual nodes in a cluster with each having a decision-making capability for events in that cluster, or the assignation of a master node that delivers instructions to neighboring nodes. Regardless of the model, the intention is leagues beyond surveillance, as it will essentially allow for 'remote control' of our surroundings. David Culler, at the University of California, Berkeley, perhaps has best described

▲ **Figure 5-17** Wireless sensor network. Top image shows prototype for a distributed sensor pod (NASA JPL). Bottom image shows 'smart dust' that functions as chemical sensors at the nano-scale level (Frederique Cunin, UCSD)

the intent of smart sensor networks: 'just as the Internet allows computers to tap digital information no matter where it is stored, sensor networks will expand people's ability to remotely interact with the physical world.'[1]

5.5 Input/Output models

At this point it is useful to summarize the different input/output models for the range of different approaches that have been discussed. In more complex sensor/actuator systems not involving smart materials, the first model discussed was the *direct mechatronic* model. Here a sensor responds to external stimulus field, its output signal is then normally transduced, and there is a direct response by the actuator. The actuator response is controlled directly by the sensed inputs. In the direct mechatronic model, sensors, transducers and actuators are normally discrete components (others may exist as well, e.g., transmitters, receivers). The *enhanced mechatronic* model next discussed contains the same features as the basic mechatronic model but now contains a logic component normally reflected through a computational environment of one type or another. Output responses are controlled by sensed inputs but can now be logically manipulated or controlled. Note that this model can become quite sophisticated. The introduction of a logic controller allows great control over actions. This kind of environment even allows the introduction of 'learned behaviors', which in turn opens up the world of complex robotic actions, artificial intelligence and other sophisticated approaches.

When *smart materials* are introduced, significant changes occur. Here we see that the property-changing characteristic of Type 1 smart materials means that the response itself is dependent upon *both* the input stimuli *and* the properties of the material. The output response remains direct, resulting in a *constitutive model I*. When energy-exchanging smart materials are used, the sensor/transducer function of the basic mechatronic model are combined into a single internal action, and, in some cases, the whole sensor/transducer/actuator function is combined into a single function. An enhanced constitutive environment, *or constitutive model II*, results that is based on energy-exchanging materials. It would normally include a logic function, which in most current applications would remain external to the material itself but would nonetheless control output responses.

So what might all of these advances ultimately aspire to achieve? Perhaps an end aspiration might well be an emulation of the biological model itself. We have briefly characterized the basic human sensory environment, and the actuation capabilities of humans and other biological forms are obvious. The 'center box' for a biological model now becomes the neurological system itself. Here there is complete internalization of all functions and logic controls into the ultimate seamless or one-part entity. Clearly, we are an enormous distance away from this model, but the aspiration level is interesting. Perhaps the basic question is that of the following: if the biological model represents the basic aspiration, why should we approach it via an emulation approach that fundamentally utilizes non-living materials?

Perhaps attention should be turned to the fascinating world that is beginning to address how living organisms can be literally designed to provide certain common needs such as motors and other functions, and ultimately even much, much more – and all within the context of a truly intelligent system. We will return to this line of thought in Chapter 8.

Notes and references

1 Cited from Huang, G.T. (2003) 'Casting the wireless sensor net', *Technology Review*, 106/6. David Culler is a member of the Berkeley research group that first coined the term 'smart dust'.

6 Smart products

For some time smart materials found their primary use in interesting but specialized engineering and scientific applications, or, at the other end of the spectrum, in novelty applications (e.g., the endless numbers of thermochromic coffee cups that change colors when filled). Recently, a whole host of new products have found their way into the market – some interesting, some not – as designers began to 'discover' them. This chapter briefly looks at smart materials from a product-oriented perspective.

In order to understand smart material use in a product context, we must first step back and look at smart materials from a broad phenomenological perspective. What do they actually do in terms that are of interest to a designer? What effects or actions are needed or wanted? After this review, we will look more directly at the results of dramatic changes that have occurred in production technologies that have ushered in new product forms or made existing ones less expensive. For example, there have been many amazing technological improvements in the production of thin films. Many exciting smart products have become possible not so much because of innovations at the basic materials level, but rather because of improved manufacturing technologies. Many of the production technologies developed have allowed many smart materials that were heretofore only laboratory curiosities to become usable to the design community.

This chapter thus focuses on identifying the actions and effects that are possible via smart materials, and then on the technologies that allow them to be implemented. The marriage between these two streams has indeed become a happy one.

6.1 A phenomenological perspective

As we have seen, most smart materials actually work at the micro-scale (smaller than a micron) and are thus not visible to the human eye. Nevertheless, the *effects* produced by these mechanisms are often at the meso-scale (approximately a centimeter) and macro-scale (larger than a meter). Whereas the physical mechanism – how the material works – is entirely dependent upon the material composition; the phenomenological effects – the results produced by the action of the material – are determined by many things independent of the material composition – including quantity, assembly construction, position and environment. As a result, very similar effects can often be produced from seemingly dissimilar materials.

We can categorize these effects in terms of their arena of action, which could be considered as analogous to an architect's intention – what do we want the material to do? The smart materials that we use can produce direct effects on the energy environments (luminous, thermal and acoustic), or they can produce indirect effects on systems (energy generation, mechanical equipment). The following categories broadly organize smart materials according to their effects that are of direct interest to designers. Note that some materials can be deployed to have multiple effects depending on the energy input.

LUMINOUS ENVIRONMENT

Transparency and color change

This is one of the largest classes of smart materials, as many different mechanisms give rise to a wide variety of color conditions. Color is understood by the human eye in two ways – by the spectral composition of transmitted light through a translucent surface to the viewer, or by the spectral composition of reflected light from a surface to the viewer. Translucent materials may change their total transmissivity, whether from opaque to transparent. Suspended particle and electrochromic technologies do this, as well as photochromics and thermotropics. Alternatively, they may selectively change the color that is being transmitted (liquid crystal, chemochromic). Reflectivities may also be changed, from one color to another (also photochromic and chemochromic) or through several colors depending on the environmental inputs (thermochromic). In glasses and films, reflected or transmitted colors may change according to the angle of view (diochroic effects). Various light control objectives, e.g., glare reduction, can also be achieved through various high performance optical materials.

Light emission

The conventional means for producing light depend upon inefficiency in energy exchange: incandescent light is produced when a current meets resistance in a wire (thereby producing infrared radiation), and fluorescent devices depend upon the resistance of a gas (thereby producing ultraviolet radiation). Light emission from smart materials is based on wholly different mechanisms, and thus is not only more efficient, but more divisible and controllable. Light can be produced of any color (electroluminescent, light-emitting diodes), of any size, intensity or shape (also electroluminescent). Light can be produced in direct response to environmental conditions (chemoluminescent, photoluminescent) and light can also be stored and re-released at a later time (photoluminescent).

THERMAL ENVIRONMENT

Heat transfer

The conventional means for adding heat or removing heat from an interior space is through a process known as dilution – air at a particular enthalpy is mixed in with the room air to 'dilute' it to the desired conditions. This is an extraordinarily inefficient process and it involves several levels of heat exchange. The most efficient heat exchange takes place in a heat engine, which essentially cycles between a low temperature and high temperature reservoir. A heat engine can be established across a junction in a semiconductor, producing an enormous temperature difference (thermoelectric). This temperature difference can be used as a sink (for cooling) or as a source (for heating).

Heat absorption

Rather than removing heat from a space (heat transfer), we can convert it into internal energy (which involves a molecular or microstructure change). Thermal energy can be absorbed and inertial swings dampened by material property changes (phase change materials, polymer gels, thermotropics).

ACOUSTIC ENVIRONMENT

Sound absorption

For many years, we have depended solely upon architectural surfaces to absorb sound, an often unwieldy enterprise as the absorption is directly proportional to the surface area. The primary method was through friction, which basically reduced the elastic energy of the sound. Energy-exchanging materials allow for more controllable and much more efficient exchange of elastic energy to another form. Elastic energy can be converted to electricity, thus reducing the amplitude of the acoustic vibrations (piezoelectric).

KINETIC ENVIRONMENT

Energy production

Although all of the energy-exchanging materials could be considered as energy producers in that they output some form of energy, we can distinguish this category according to the purpose of that energy. If the output energy is intended for some other function, such as sensing, then we do not consider the material typologically to be an energy-producing smart material (light emission is an exception as it has its own distinct category). The materials in this category are those that we can consider as 'generators' – they directly produce useful energy. The energy can be in many forms: generated electricity (photovoltaic and thermo-photovoltaic), heat pump or engine (thermoelectric) as well as elastic energy (piezoelectric).

Energy absorption (mechanical dampening)

The counterpoint to the energy-producing materials, whose focus is the form of the output energy, are materials that focus on the form of the input energy. More precisely, the intention of energy-absorbing materials is to dissipate or counteract the input energy. Vibrations can be dissipated by conversion to electricity (piezoelectric) or dampened by absorption produced by a material property change (magnetorheological, electrorheological, shape memory alloy). Column buckling can be counteracted by an applied strain (piezoelectric) and other types of deformations can also be counteracted by selectively applied strains (electrostrictive, magnetostrictive, shape memory alloys).

Shape change

Unlike color change, which can take place over a large area of material, shape change tends to be confined to a much smaller scale (typically meso-scale). This is due to inherent limitations in the scaling of dynamic forces. Nevertheless, even though all materials undergo some form of shape change from an energy input (i.e. the elongation of a metal rod under tension, the swelling of wood when saturated with water), the shape-changing smart materials are differentiated by not only their ability to be reversible, but also by the relative magnitude of the shape change. For example, smart polymer gels (chemotropic, thermotropic, electrotropic) can swell or shrink volume by a factor of 1000. Most shape-changing materials move from one position to another – the movement may be produced by a strain, or it may be due to a microstructural change – but the result is a displacement. A material may bend or straighten (shape memory alloys, electrostrictive, piezoelectric), or twist and untwist (shape memory alloys), or constrict and loosen (magnetostrictive), or swell and shrink (polymer gels).

6.2 Product technologies and forms

The phenomenological perspective just presented emphasizes the immediately tangible results of the actions of smart materials. In order to accomplish these ends in real products or devices that are targeted for specific applications or uses, it is necessary that smart materials be made available in forms useful to the designer, e.g., filaments, paints or films. At the beginning of the chapter, we noted that the recent explosion in the use of smart materials has been engendered to a large part by new developments in manufacturing technologies – especially for polymers. For example, new film technologies for polymers have enabled various forms of view-directional films to become basic product forms that can subsequently be used in higher-level products targeted for use in either product design applications – e.g., privacy screens for computer displays – or architecture – e.g., privacy windows. In the following, we will first briefly look at processes for making polymeric films and other materials because of their fundamental importance to the development of smart product forms. We will also look at several key technologies – notably thin film deposition processes – that underlie how many smart products actually work. A look at general smart product forms available to the designer, e.g., paints, films, glasses, dyes, will follow.

BASIC PROCESSES FOR POLYMERIC PRODUCTS

Improved processes for making polymeric products – filaments, strands, films – have particularly had a profound impact on making some smart materials ubiquitously available – notably processes for making polymer films and strands and processes for depositing thin layers of different materials on various substrates.

All forms of polymeric material (filaments, sheets) must be drawn in order to achieve the necessary long chain molecular structures that characterize useful polymers (see Chapter 2). Actual processes used depend on the basic material itself, its intended product form, and whether special properties are to be imparted to it. Common thin sheet materials are made by an extrusion process. Granulates or powders of base material are heated and mixed, and then extruded under high pressure through a slit-die. The hot emerging sheet is pulled through a cooling cycle onto a roll. Various additional processes can be employed for biaxial stretching to specially orient molecular structures. The resulting optical properties may be controlled by adjusting process means and factors. Filaments are similarly drawn. For three-dimensional shapes, various film-blowing (blow-molding) techniques are used. Casting techniques are occasionally used for special shapes. Blow and cast films have low levels of molecular orientation and are thus weak in tension but can have strong tear resistance. Various processes exist for surface texturing, printing, or adding coatings (including metallic coatings).

DEPOSITION PROCESSES

The importance of surface-related phenomena in many behaviors has been repeatedly stressed. The color of an object, for instance, depends very much on its surface characteristics. It follows that direct interventions at the surface level can have profound effects. Consequently, many basic technologies have been directed toward this end. In particular, 'thin film deposition processes' as they are commonly called have been developed for adding thin layers of different substances to basic substrates (the term 'thin layer deposition processes' is actually preferable to avoid confusion with 'film' as a product form). Paintings and coatings, discussed below, are much thicker than these micro-level layers.

Basic deposition processes, including the time-honored 'electroplating' process known to most, have been around for a long time. Recently, however, a number of new processes have been developed that allow depositions to be made on a variety of substrates. Substrates that can be used include glass, metals, polymer film and many others. There are currently many different kinds of deposition technologies for material formation, including evaporation, plasma-assisted processes (including sputter deposition), chemical vapor deposition and others. Many current devices are quite sophisticated. The sputtering device in Figure 6–2 involves molecules being literally drawn from a source and deposited on a substrate in a controlled way. Extremely thin films (some can even be at the nanometer level) can be formed.

Development of these deposition processes has been driven by electrical engineers and others because of their value in different kinds of microelectronic solid-state devices. Depositions need not be in single layers only. In the

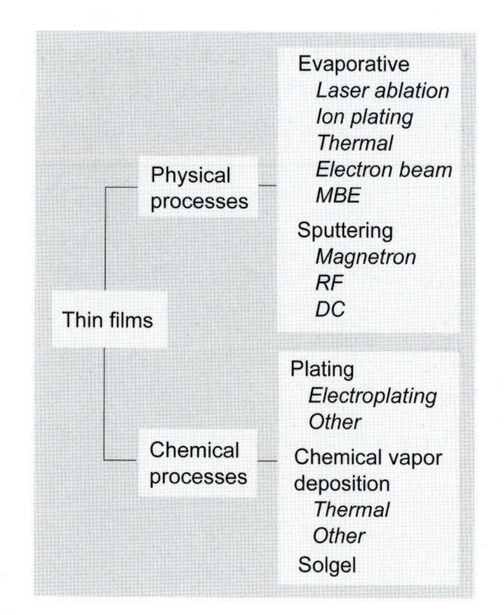

▲ **Figure 6-1** Common methods for creating thin films or layers on other materials

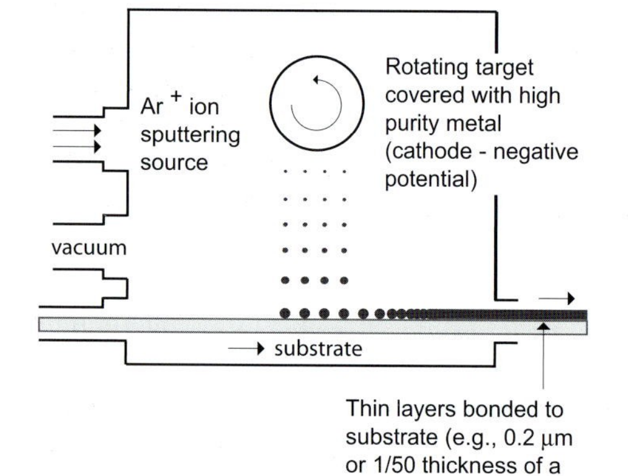

▲ **Figure 6-2** A sputtering device for depositing a thin layer of material on a substrate

microelectronic world, electrically conductive structures can be created by using these techniques. Various oxides can be deposited. An insulating or dielectric layer can also be deposited to create multi-level structures. Sequences of different layers may be repeated many times. Various masking techniques can be used to make lines, openings, patterns and so forth. Simpler single-layer depositions can be applied to large areas.

6.3 Smart material product forms

POLYMER FILMS

Polymeric films are seemingly everywhere – including the shiny holiday season gift wrapping that changes colors in different light. Technologies for making films from different materials have been around a long time. Recently, there have been new developments that yield even thinner and tougher films that can be designed to have many different properties and exhibit a wide variety of different behaviors. Particularly interesting are developments in the area of multi-player laminates. These products can be relatively low cost. Several are described below. Many are actually high-performance materials, while others exhibit true smart behaviors.

Radiant color and mirror film

The 3M™ Corporation has developed and actively promoted many types of high performance films, including radiant mirror film and radiant color film. These are remarkably interesting products. The mirror film is advertised as specularly reflecting 98% of visible light, which makes it attractive for a wide variety of applications. The opaque mirror film consists of multiple layers of polymeric film, each with differing reflective properties, and a polyester surface. It can be embossed, cut, coated to be UV resistant and given an adhesive backing or laminated to other surfaces. It is metal free (hence non-corrosive) and thermally stable.

Radiant color film also consists of multiple layers of film with different reflective properties, but is transparent. It possesses remarkable reflective and transmissive qualities. The color of the reflection perceived depends on the angle that light strikes it. The color of the film when looking through it depends on the exact angle of the viewer to the film surface. These qualities, similar to more expensive diochroic glass, render it an immediately fascinating material to the designer. Many of these properties are illustrated in the accompanying figures.

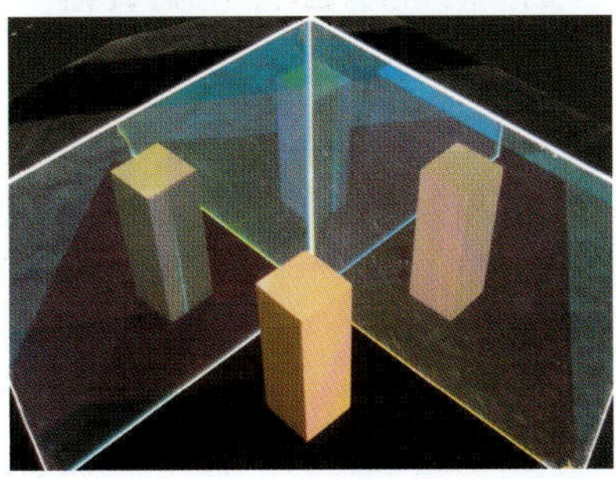

▲ **Figure 6-3** Radiant color film. The same block is shown from different vantage points

Since these films consist of multiple layers, it is possible to create different kinds of films with different optical properties fairly easily. Transmission levels can be varied, as can the spectral response to different wavelengths.

▲ **Figure 6-4** Radiant mirror film. The color of the transparent film depends on the angle of the viewer with respect to the film

View directional films

Often called light control film or privacy film, this polymeric material is embossed with very small specially shaped grooves or micro-louvers. Depending on how the micro-louvers are organized, a viewer can see through the film only in specified directions. As the viewer changes locations, the film becomes less and less transparent. The observer thus perceives an object through the film only under certain conditions. Films with different view angles for transparency, e.g., straight on or at some angle, can be obtained.

The interesting characteristics of this film have led to its use on store fronts and many kinds of displays. Imagine walking on a catwalk with the film beneath you. The pathway ahead of you and behind you looks solid, but beneath you it is transparent. This same film can also be effectively used to control light coming in through it, and thus it finds wide application as a glare-reducer for computer displays and other applications.

Image redirection films

While seemingly similar to view directional films, these view redirection films have the curious property of acting somewhat like a periscope in that one can look around corners to a certain extent. They are made from embossing specially shaped grooves onto polymer sheets.

▲ **Figure 6-5** View control film (privacy film) allows the viewer to see an object clearly only from a specified direction. As the angle of view changes, so does the visibility of the object

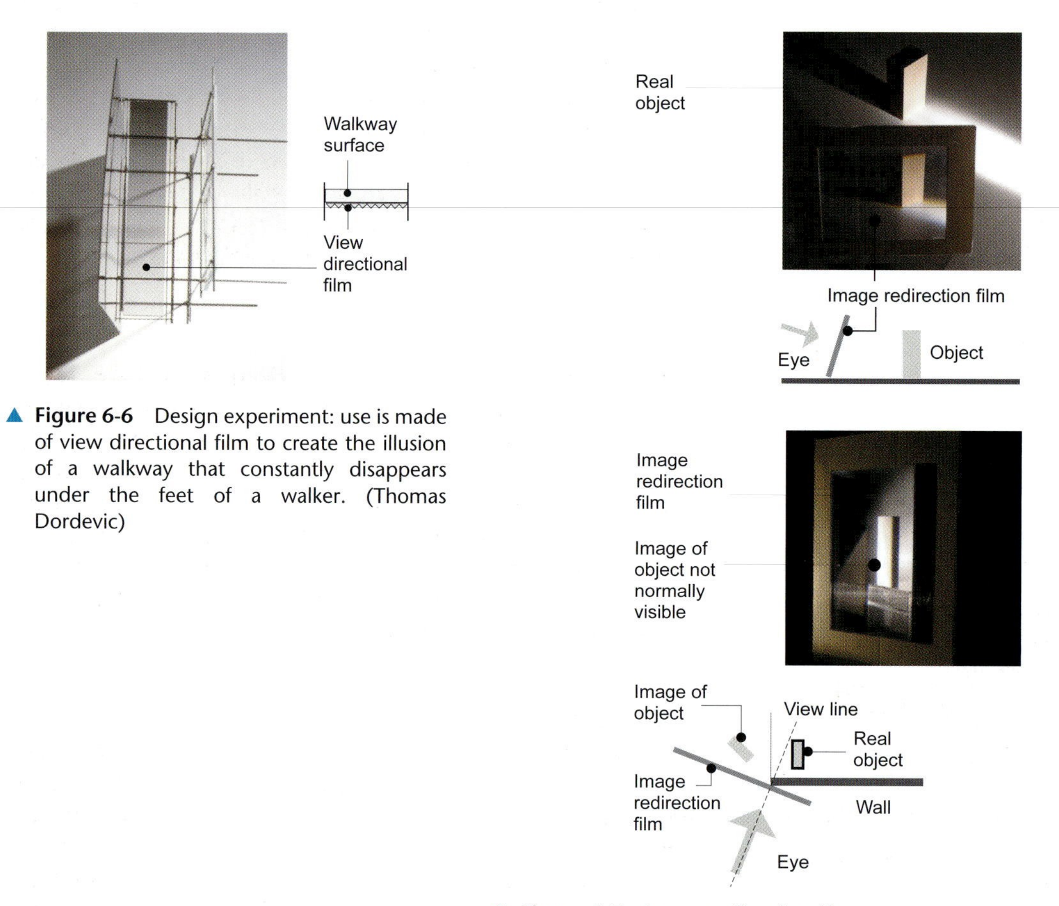

▲ **Figure 6-6** Design experiment: use is made of view directional film to create the illusion of a walkway that constantly disappears under the feet of a walker. (Thomas Dordevic)

▲ **Figure 6-7** Image redirection film

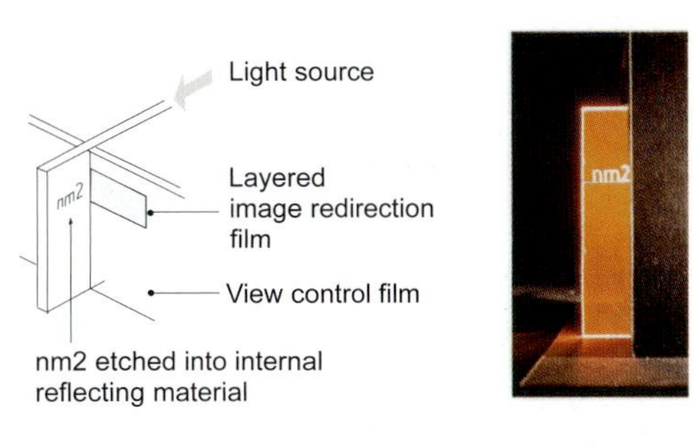

▲ **Figure 6-8** Design experiment: as the angle changes, images of the bright nm2 sign that is lighted by internal reflection are seen first in the view control film, and then in a much larger way in the image redirection film. (Jonathan Kurtz)

Fresnel lens films

The Fresnel lens has achieved wide use since its inception by August Fresnel in the 18th century because of its ability to focus parallel light rays on a point, or to be a highly effective means of projecting bright parallel rays from a point source (this latter ability made the use of the Fresnel lens widespread in lighthouses). Original lenses were made of high quality optical glass. The necessary lens shaping is now possible on thin polymer films that can be inexpensively produced. They are now widely used in many applications, ranging from overhead projectors to campers' solar cookstoves.

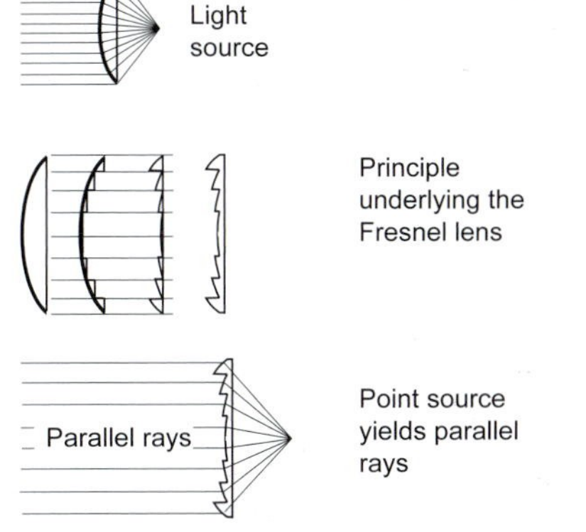

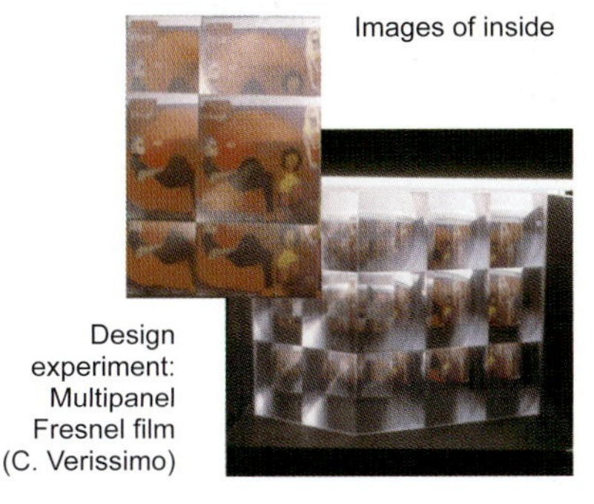

▲ **Figure 6-9** Fresnel lens

Polarizing films

The interesting properties of polarized light have been explored in Chapter 3. The advent of new ways of making polymeric films has led to the development of relatively inexpensive polarizing films. Some of these films have adhesive backings that allow them to be applied to glass substrates. Many kinds of polarized films are used in computer or kiosk displays to reduce glare. A circularly polarized film assembly consisting of linear and circular polarizing filters can be particularly effective here.

Light pipes

Many kinds of polymer films with special surface properties are shaped into tubes to be used as devices for transmitting light. Several different varieties are available with varying degrees of efficiency. Some are designed to 'leak' light along their lengths to create glowing tubes. Others are designed to carry light with as little loss as possible from one end of a tube to another.

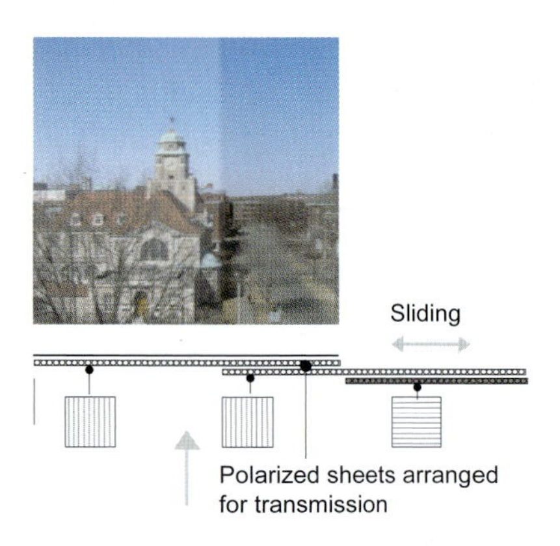

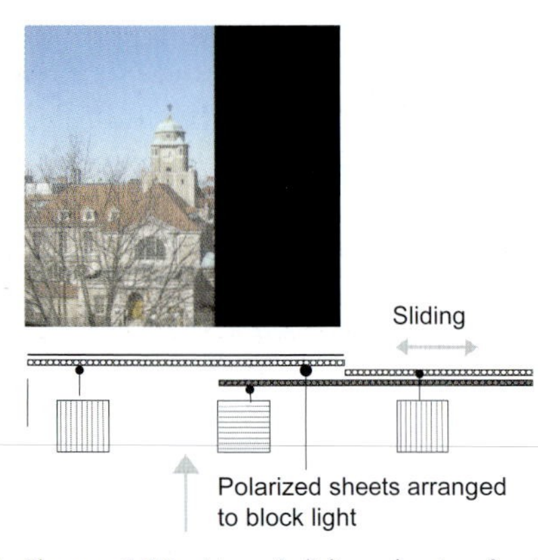

▲ **Figure 6-10** Use of sliding sheets of polarized film to modify a view

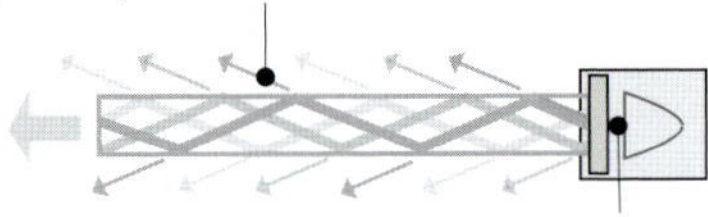

▲ **Figure 6-11** Light pipes work by reflecting light along the inside of a tube. A portion of the light escapes along its length to create a bright tube

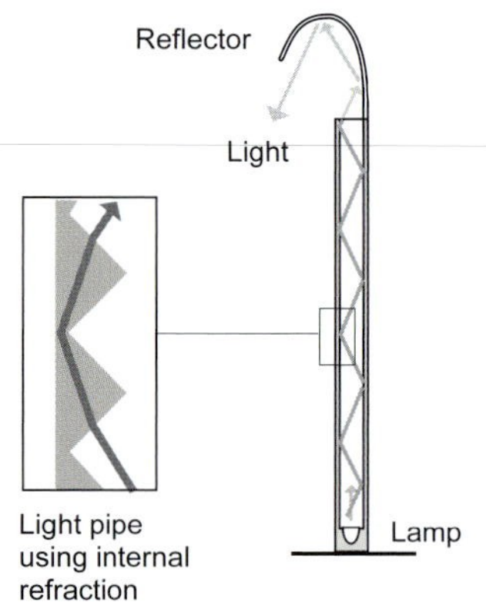

▲ **Figure 6-12** External lighting fixture that uses a refraction-based light pipe. This arrangement allows for improved light distribution and easy maintenance and replacement of lamps

Photochromic films

Photochromic materials change color when subjected to light. Many photochromic films are available that change from a clear state to a transparent colored state. These polymeric films can be relatively inexpensive as compared to photochromic glasses. Normally, their color-changing response is relatively slow and the color quality less controlled than obtainable in photochromic glasses.

Thermochromic films

Thermochromic materials change color with temperature. Special thermochromic films, based on a form of liquid crystal behavior, can exhibit controlled responses to temperature changes. They can be designed to be calibrated to specific temperature ranges. The common 'thermometer strip' for measuring a human's body temperature via a color-coded thermochromic film is carefully calibrated.

Electroluminescent films

Electroluminescent materials, described in Chapter 4 produce illumination when their phosphor materials are charged. This phosphorescent material can be put on a film layer, as can metallic charge carriers. This technology is directed towards thin low-voltage displays with low power consumption. It is largely compatible with a number of low-cost fabrication techniques for applying it to substrates (e.g., spin coatings) and other printer-based fabrication techniques. For a while, these films were considered an exciting possibility for large-scale lighting; but interest in them waned because of the development of light-emitting diode (LED) technologies.

Conductive polymeric films

The idea of polymeric materials conducting electricity is a seemingly new and exotic one. Forms of conductive polymers have, however, been in wide use for a long time. These common conductive polymers are normally called 'filled polymers' and are made by adding to the polymer a conductive material such as graphite, metallic oxide particles, or other conductors. The addition of fillers is easy in many polymeric materials, particularly thermoset plastics such as epoxies. Doing so in thermoplastics that come in sheet form is more difficult. Deposition processes can also be used to directly give polymeric films a conductive coating. Ink-jet printing processes using metallic materials can be used as well, particularly for specific patterns.

As discussed in Chapter 4, conjugated polymers based on organic compounds can be directly conductive. For polymers, the materials used are usually based on polyaniline or polypyrrole compounds. At the molecular level they have an extended orbit system that allows electrons to move freely from one end of the polymer to the other end. These inherently conducting polymers are also sensitive to radiation, which can change the color and the conductivity of polyaniline.

These materials are widely used in organic light-emitting polymer (OLEP) films (see below). Additionally, different electronic components like resistors, capacitors, diodes and transistors can be made by combining different types of conducting polymers. Printed polymer electronics has attracted a lot of attention because of its potential as a low-cost means to realize different applications like thin flexible displays and smart labels. A form of electronic paper has been proposed based on these technologies.

These electroactive polymers can also be used as sensors, actuators and even artificial muscles. An applied voltage can cause the polymer to expand, contract or bend. The resulting motion can be quite smooth and lifelike. The motions demand no mechanical contrivances, and are thus often compared to muscles – hence the term 'artificial muscle'. There have been interesting experiments, for example, with these polymers in trying to replicate fish-like swimming motions. Developing, controlling and getting enough force out of these materials to really act like artificial muscles has always been problematic. Until recently, electroactive polymers have presented practical problems. They consumed too much energy. They couldn't generate enough force. Alternatively, bending them could generate voltages (see piezoelectric films below) which makes them useful as sensors.

Light-emitting polymers

There are several technologies based on polymeric materials that emit light. There has been great interest in this area because of the potential for low costs, their ability to cover large areas and their potential for material flexibility. Electrically conducting or semiconducting organic polymers have been known since the beginning of the 1990s when it was observed that some semiconducting organic polymers show electroluminescence when used between positive and negative electrode layers. This led directly to the development of organic light emitting diodes (OLED) and films. The polymer light emitting diode (PLED) is made of an optically transparent anode metal oxide layer (typically indium tin oxide or ITO) on a transparent substrate, a layer of emissive polymer (such as polyphenylene-vinylene), and a metal cathode layer. Typically, the metal cathode layer is based on aluminum or magnesium and is evaporated onto the organic layer via vacuum metal vapor deposition techniques. An applied voltage causes the sandwiched emissive polymer to emit light. The chemical structure of the polymer can be varied so that the color of the light can be changed. Necessary voltages are low.

Photovoltaic films

The basic photovoltaic effect was discussed in Chapter 4 and is again explored in detail in Chapter 7. Of interest here is that flexible polymeric films of exhibiting photovoltaic effects have been made as a result of advances in laminating multi-layered films. Specific ways of making films vary. Some approaches are based on the *p–n* effect and use a mix of polycrystalline compounds (e.g., gallium, copper, indium, gallium and selenium). They are grown by a co-evaporation process on a film (see below) and assembled into a multi-layer structure, normally with a metallic back contact and a conducting, radiation-transmitting front layer. Another approach uses solid state composites of polymer/fullerene compounds. A layer is made of special carbon molecules called fullerenes that have high electron affinities. This layer draws electrons from another layer of a positively charged polymer that can be photo-excited. A current is created between the negatively charged fullerenes and the positively charged polymer.

The objectives often stated by developers are to create thin and flexible solar cells that can be applied to large surfaces, and which could be made in different transparencies and colors so that they could be used in windows and other similar places. Problems of low efficiency, including those generated by not being able to control solar angles in these applications, remain. Heat build-up and energy conversion problems are also fundamental issues. There have been, however, many successful applications in the product and industrial design worlds for smaller and more contained products, ranging from clocks to battery chargers.

Piezoelectric films

Piezoelectric materials convert mechanical energy (via deformations) to electrical energy and vice-versa (see Chapter 4). Piezoelectric films have been developed that are based on polarized fluoropolymers (polyvinylidene fluoride – PVDF). It comes in a thin, lightweight form that can be glued to other surfaces. The film is relatively weak as an electromechanical transmitter compared to other piezo forms. Large displacements or forces cannot really be generated. These films can be used, however, as sensors to detect micro-deformations of a surface. Hence they find use in everything from switches to music pickups. The same PVDF material also exhibits *pyroelectric* properties in which an electrical charge is produced in response to a temperature variation.

Chemically sensitive color- and shape-changing films

Films have been developed that are sensitized to respond to different chemical substances that act as external stimuli. Exposed films may changes shape, color or other properties. Interest in these films has been widespread because of their potential in acting as simple sensors that detect the presence of chemicals in surrounding atmospheres or fluids. An interesting further development for shape-changing polymers is to couple them with holographic images. The holographic image presented to the user could thus change as a function of the swelling or contraction of the film. Hence, different 'messages' or other information content could be conveyed.

Other films

A whole host of other films have been developed that can be used independently or added to different substrates. In many cases films are coated in some way to provide specific properties; in other cases they are made up of many laminated layers with different properties. *Antireflective films* seek to reduce reflection or glare and to improve viewing contrast. They are widely used for electronic displays but have found use in architectural settings as well. *Brightness enhancement* films have been developed with the intent of increasing the brightness of computer displays. They do this by focusing light towards the user. *Holographically patterned* films have metallized coatings that can hold holographic images and can thus be used to transmit previously inscribed lighting patterns (see discussion below under holographically patterned glasses). Many other films are available as well.

POLYMER RODS AND STRANDS

Optical carriers

There are many types of optical cables, rods or fibers available for use in transmitting light. Glass is widely used as a carrier material because it has very low attenuation or light loss over its length. However, glass is relatively expensive, difficult to cut and requires special end connections. For many applications, various kinds of plastic rods and strands can be used instead of glass. Plastics are relatively inexpensive and easier to cut and connect than glass. Plastics are normally used in only short distance applications and where attenuation losses are not significant. Consequently, plastics find wide usage in lighting systems.

Optical cables can also be made in many different ways. At the most basic level, simple long flexible plastic strands or rods find uses in many simple applications that involve simple light distribution via internal reflection (see Chapter 3). These same rods can be encased or jacketed in an opaque material to improve their light transmission. Diameters can vary greatly, but even large diameters suitable for lighting installations can be relatively inexpensive.

In more demanding uses, more complex arrangements are used. A true fiber-optic cable generally consists of a layered system with an inner core of optically transparent material that transmits light. This core is surrounded by an outer covering of another optically transparent material, but one with a lower refractive index than the inner core. A surrounding outer jacket encases both the core and its cladding for protection. Different internal arrangements of core/cladding components are possible depending on the application and cost constraints. Core and cladding materials can be made of polymeric materials. For example, a core of polymethyl methacrylate polymer (PMMA), cladding of a fluorine polymer, and a polyethylene jacket is often used.

Shape-changing polymer strands

These materials hold promise for a great number of applications. Polymers that shrink or expand due to changes in the thermal environment, for example, have been explored for use in the surgical field. Inserted around blood vessels, body heat causes them to literally tie themselves into a remembered knot.

INKS AND DYES

Smart dyes and inks are fundamental to the making of many types of smart products, including papers, cloths and others. Dyes come in highly concentrated form and can be used as a basis for transforming many common materials into 'smart' materials. Normal paper, for example, can be made into thermochromic paper by the use of leucodyes. When cool, leucodyes exhibit color and become clearer upon heating or can be made to change to another color. Photochromic dyes can be used to make photochromic cloths. Color-changing printing can be done via thermochromic or photochromic inks. Applications of smart inks are widespread since they can be used with most major printing processes, including offset lithography, flexography and so forth.

SMART PAINTS AND COATINGS

Painting and coatings are ancient techniques for changing or improving the characteristics or performance of a material.

The development of smart paints and coatings gives these old approaches new capabilities. Smart paints and coatings can be generally classified into (a) high-performance materials, (b) property-changing materials and (c) energy-exchanging materials. In today's world there are so many specially developed high-performance paints and coatings – particularly those that are the result of the burgeoning field of polymer science – that any detailed coverage is beyond the scope of this book. Here we will concentrate on those paints and coatings that are developed with the specific intent of being 'smart'.

By way of definition, paints are made up of pigments, binders and some type of liquid that lowers the viscosity of the mixture so that it can be applied by spreading or spraying. The pigments may be insoluble or soluble finely dispersed particles, the binder forms surface films. The liquid may be volatile or nonvolatile, but does not normally become part of the dried material. Coatings are a more generic term than paints and refer to a thicker layer. Many coatings are nonvolatile.

As with many other applications, many of the basic property-changing materials discussed earlier can be manufactured in the form of fine particles that can be used as pigment materials in paints. Thus, there are many variations of *thermochromic* and *photochromic* paints or coatings. Thermochromic paints are widely used to provide a color-change indicator of the temperature level of a product. Special attention must obviously be paid to the chemical nature of the binders and liquids used in formulating paints of this type so that the property-changing aspects of the pigment materials are not changed. These same chromic materials still often degrade over time, particularly when exposed to ultraviolet radiation.

Other property-changing materials could be incorporated into paints and coatings as well, but the value of doing so must be carefully considered. Some phase-changing materials, for example, could be directly used in coatings or embedded as microcapsules. Whether or not sufficient amounts of the material could be incorporated to achieve the thermal capabilities desired in a usable product, however, is another matter.

In the sphere of energy-exchanging materials used in paint or coating form there are many direct applications. There are many natural and synthetic luminescent materials that can be made in paint or coating form. These paints or coatings absorb energy from light, chemical or thermal sources and re-emit photons to cause fluorescence, phosphorescence or afterglow lighting (see Chapter 4). Again, care must be taken with the chemical natures of the binders and liquids used in conjunction with these materials.

Many paints and coatings are devised to conduct electricity, such as the coatings used on glass substrates to make the surface electrically conductive and thus have the capability of 'heating up'. The advent of *conducting polymers* (see above and Chapter 4) has opened a whole new arena of future development for paints and coatings since paints and coatings have often been polymeric to begin with. The possibility of these paints and coatings now being electrically conductive is interesting. Potential applications vary. There has been a lot of recent interest in making smart paints that can detect penetrations or scratches within it, or corrosion on the base material. A heavy scratch, for example, would necessarily change the associated electrical field, which could in turn possibly be picked up by sensors.

Polymeric materials can also be used as hosts for many other energy-exchanging materials, including *piezoelectric particles* (recall that piezoelectric materials produce an electrical charge when subjected to a force, or can produce a force when subjected to a voltage). Coatings based on these technologies are being explored in connection with 'structural health' monitoring (see Chapter 7). Deformations in the base material cause expansions or contractions in the piezoelectric particles in the coating that in turn generate detectable electrical signals. These electrical signals can be subsequently interpreted in many ways to assess deformation levels in the surface of the coated materials. Assessing directions of the surface deformations that produce the measured voltages, however, remains difficult. These same technologies can be used to evaluate the vibration characteristics of an element, including its natural frequencies.

In these smart piezoelectric paints, piezoelectric ceramic particles made of PZT (lead ziconate titanate) or barium titanate ($BaTIO_3$) are frequently used. They are dispersed in an epoxy, acrylic, or alkyd base. The paint itself is electrically insulating and, in order for the paint to work as described, an electrode must be present (on the film surface) to detect a voltage output. Measurements can be obtained only in the region of the electrode. Arrays of electrodes, however, may be used with data obtained from each to yield a picture of the behavior of a larger surface. In large applications, simple electrodes may be made by using electrically conductive paint applied over the piezoelectric. Thin lead wires to these 'painted electrodes' are needed and may in turn be covered by a coating. Other more sophisticated ways of making more precise electrodes are also in use. These interesting applications are, by and large, still in the research and development stage.

GLASSES

Electro-optical glass

Electro-optical glass is a good example of a successful application of thin film technology in a design context. Glass is well known for use as an electrical insulator. As a dielectric material, it inherently does not conduct electricity. This very property that is so advantageous for many applications, however, becomes problematic for other applications – especially in this world of flat panel displays and other technologies that could seemingly effectively use glass for other purposes than as simply a covering material.

Electro-optical glass has been developed with these new needs in mind. Electro-optical glass consists of a glass substrate that has been covered – via a chemical deposition process – by a thin and transparent coating of an electrically conductive material. The most frequently used product uses a chemical vapor deposition system to apply a thin coating of tin oxide to a glass substrate. The chemical deposition process yields a coating that is extremely thin and visually transparent, but which is still electrically conductive.

In architecture, this technology can be used to create 'heated glass'. Strip connectors are applied to either edge of a glass sheet and a voltage applied. The thin conductive deposition layer essentially becomes a resistor that heats up. The whole glass sheet can become warm. The potential uses of heat glass of this type in architecture are obvious. Difficulties include finding ways to distribute the current uniformly over the surface.

Dichroic glass

A *dichroic material* exhibits color changes to the viewer as a function of either the angle of incident light or the angle of the viewer. The varying color changes can be very striking and unexpected. Similar visual effects have long been seen in the iridescent wings of dragonflies and in certain bird feathers; or in oil films on water surfaces. Recent innovations in thin layer deposition techniques have been employed to produce coatings on glasses to exhibit dichroic characteristics.

In dichroic glass, certain color wavelengths – those seen as a reflection to the viewer – are reflected away while others are absorbed and seen as transmitted light. The colors perceived change with light direction and view angle. The dichroic

▲ **Figure 6-13** 'Diochroic Light Field' – an installation by James Carpenter, New York City

(originally referring to two-color) effect has been technically understood for many years. In new dichroic glass, a glass substrate is coated with multiple layers of very thin transparent metal oxide coatings, each with different optical properties. When light impinges upon or is passed through these layers, various complex optical effects occur. Fundamentally, reflections are created when light passing through a layer of one optical index of refraction meets a layer with a different optical index of refraction. When multiple transparent layers are present, different reflection directions can develop at different material change points. A further effect is that the layers can become plane polarized when they absorb light vibrating in one orientation more strongly than the other. The anisotropic materials in the layers then exhibit a change in color when viewed from different directions. Interference takes place because of the multiple layers in which certain wavelengths combine with others to create new wavelengths of added or subtracted intensity and corresponding color changes. Carefully altering or controlling the properties of the different layers can achieve different color effects.

Dichroic glasses are made using thin layer deposition techniques (see previously). Materials such as magnesium, beryllium, selenium or others are used as the deposition material. Normally, electron beam evaporation and vacuum deposition processes are used. Glass to be used as the thin film substrate is normally put in a vacuum chamber and an electron beam is passed over the material to be vaporized. The vaporized material is ultimately deposited or condensed on the glass. Since uniformity of deposition is critical, rotating chambers are often used (albeit other approaches are possible). Layers are only a few millionths of an inch thick. The number of layers deposited varies, but can be as high as 30 or 40. By careful selection of materials for different layers (i.e., looking at their optical properties and thicknesses) different kinds of primary and secondary color reflection and transmission properties can be achieved. The process is quite complex, and hence dichroic glass is expensive.

Dichroic glass has been effectively used in many design situations. It is often best used selectively (see Figure 6–13, showing a dichroic light field by James Carpenter). The coatings that produce the dichroic effect are subject to abrasion; hence a protective glass layer is typically used as a protection.

Holographically patterned glass

This glass is currently used for optical and related lighting purposes. The desired optical effects (normally in the form of light patterns) are inscribed beforehand in the microstructure of the surface of the material and are essentially replayed when a light is transmitted through the material. They allow the light to be directed into particular patterns. These particular luminous distributions are recorded *a priori* holographically on a reflective metallized coating that has been applied to a glass substrate. These materials are finding increasing use as diffusers in lighting applications, since, unlike the uncontrolled light spread of conventional diffusers, these surfaces can be engineered to yield particular light spreads. These diffusers are also transparent and provide relatively distortion-free images at certain viewing distances.

Other glasses

As with films, glasses can be coated in a great number of different ways to provide specific properties; while in other cases films with different properties may be layered onto basic glass substrates. Hence, as with polymer technologies, products such as *antireflective glass* or *brightness enhancing glass* can be obtained. Glass with special thermal properties ('glass with low E coatings') can be similarly obtained. The use and quality of *photochromic glass* has been developed to a remarkable extent because of the huge market in photochromic sunglasses. Quality control and response times are excellent. Glasses are also widely used as substrates or carriers for a wide variety of other smart technology approaches (e.g. electrochromics, LCDs, suspended particles).

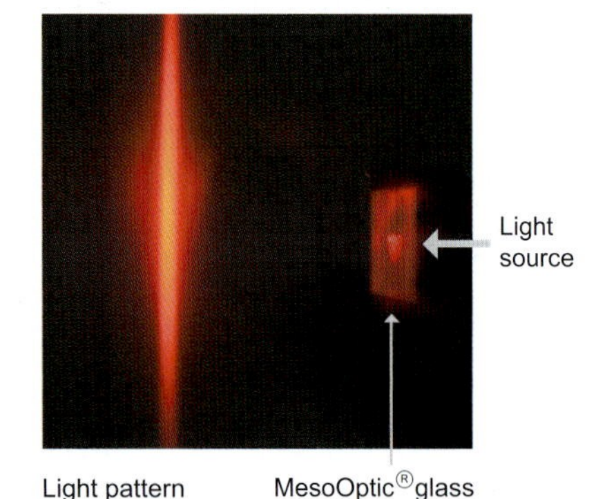

▲ **Figure 6-14** MesoOptic® glass is inscribed with a holographic image to produce a predefined light pattern.

SMART FABRICS

The term 'fabric' refers to a material that in some way resembles or shares some of the properties of cloth. At first glance, the idea of a 'smart' fabric may seem curious, but smart fabrics represent an area of enormous potential. In this discussion, we will focus our attention largely on *woven materials* and *flexible layered materials,* as there is a clear overlap with films. Many applications developed to date are for clothing, but similar technologies can be envisioned as applying to the many fabrics used in architecture or product design.

Several primary types of smart fabrics exist:

- High-performance fabrics with materials or weaves designed to accomplish some specific objective.
- Fabrics that exhibit some form of property change.
- Fabrics that provide an energy exchange function.
- Fabrics that in some way are specifically intended to act as sensors, energy distribution, or communication networks.

The first class of fabrics discussed is comprised of high-performance flexible materials and not, strictly speaking, smart materials. Many types of materials and fabrics are specifically engineered to accomplish a particular performance objective related to light, heat, acoustic properties, permeability, structural strength, etc. This is a huge class of flexible materials. Here we will look only at a few selected examples to give a sense of the field.

Light and color

There are many fabrics that deal in one way or another with light and color. Different kinds of films with special reflective or transmission qualities can also be applied to traditional fabrics or directly woven into them, imparting many of the qualities of films discussed above. Fabrics may be made of materials with different optical qualities, and thus reflect light from only certain angles. Fabrics can also be made of layers of transparent materials with different refractive indices. Depending on the layering and the materials used, these fabrics can reflect light within certain wavelengths and absorb light not within this range. A 'mirror' material can be made that reflects light in all directions with little absorption.

In addition to the use of these fabrics in displays or as special wall surfaces, fabrics that deal with light or radiation reflection and transmission also find many of their more obvious applications in goods for the sporting industry (e.g., emergency blankets).

Fiber-optic and electroluminescent weaves

The use of optical fiber-optic strands to make fabrics has opened the door to a variety of applications, including the woven fabrics that exhibit remarkable visual characteristics.

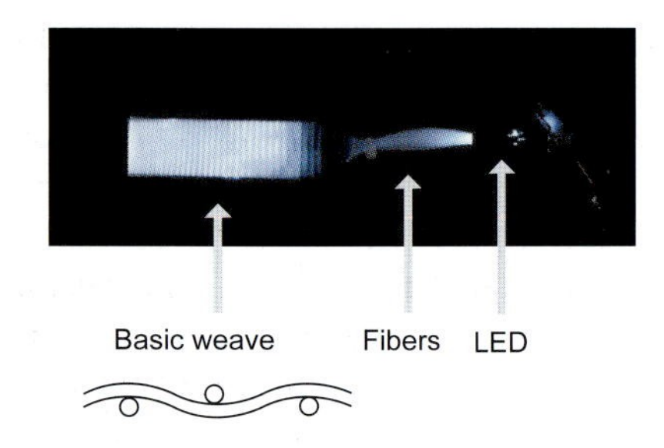

▲ **Figure 6-15** Fiber-optic weave material

They have remarkable visual appeal. One approach uses two layers of optical fiber weaves sandwiched between an outer reflective Mylar layer and an outer transparent diffuse layer. The fiber-optic weaves are in turn connected to a light source, typically an LED. Light is emitted from abraded surface areas in the fiber-optic weave. Other approaches use optical fibers in one direction only, with neutral fabrics running perpendicular to them. Other fabrics incorporate fiber-optic strands for use as sensors. A similar weaving strategy can be used in connection with electroluminescent materials.

Breathable fabrics

Another class of high-performance fabrics deals with material porosity or permeability. Of particular interest here are well-known examples such as the polymer-based membrane materials used in many sporting goods (jackets, boots) that are more or less waterproof, but still allow moisture vapor permeability for 'breathability'. Products of this type are normally based on polytetrafluoroethylene (developed in 1938 and commonly known by the DuPont brand name Teflon). The material is stretched into a porous form to form the 'breathable' membranes widely used today under brand names such as Gore-Tex®. A host of medical applications exist for these same kinds of materials (filtration systems). Since they are engineered materials, characteristics such as liquid entry pressure, biocompatibility, chemical stability and other factors can be predetermined, thus allowing a range of medical and industrial applications (e.g., filters, vents, gaskets, sensor covers, pressure venting).

Property-changing fabrics – thermochromic and photochromic cloths

The second class of smart fabrics contains those that exhibit some form of property change when subjected to an external stimulus. While many of the property-changing smart materi-

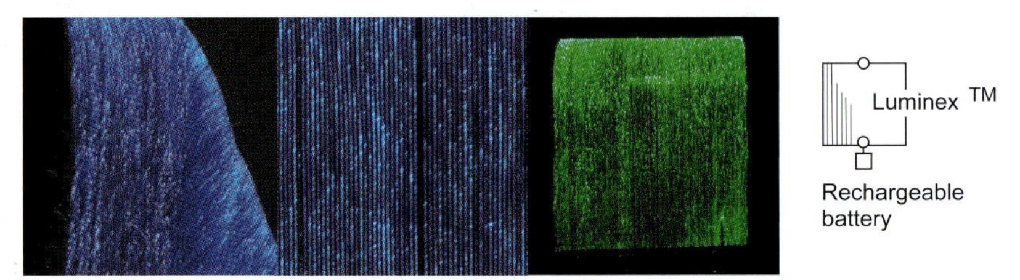

▲ **Figure 6-16** Optical cloth. The material is composed of fiber-optic strands lighted by periodically spaced LEDs. The strands are woven with other materials. (Courtesy of Luminex™)

als discussed earlier could indeed be made into fabrics, the most common applications are for fabrics that have color change properties. These are typically traditional fabrics that are either impregnated with thermochromic or photochromic materials in dye form, or are layered with similar materials in the form of coatings or paints (see above). Most commercial applications here are currently at the novelty level, e.g., color-changing T-shirts. Larger architectural applications can be easily envisioned, but remain hampered by the problem that many of these dyes and paints degrade when subjected to the ultraviolet radiation found in normal sunlight, which restricts their long-term exterior use.

Phase-changing fabrics

A number of highly interesting property-changing fabrics have been developed for controlling thermal environments. These fabrics normally incorporate phase change materials. As discussed in Chapter 4, phase change processes invariably involve absorbing, storing or releasing of large amounts of energy in the form of latent heat; and can thus be very useful in controlling thermal environments. Phase change materials have successfully been incorporated into textiles via the use of micro-encapsulation technologies. Phase-changing materials are embedded in tiny capsules and distributed through the material. The phase-changing materials within the capsules can be engineered to undergo phase changes at specified temperatures. While architectural applications can be envisioned, most commercial applications are found in outdoor sporting goods (gloves, coats, socks). Current micro-encapsulation processes are targeted for the latter applications. Other forms of encapsulation can be envisioned for architectural applications.

Outside cold environment

'Thermocules' (OUTLAST™)

Body

Excess body heat is absorbed

Heat is released when body is cool

Encapsulated phase-changing materials absorb or release heat as needed

'Thermocules' in fabric

'Thermocules' in fiber

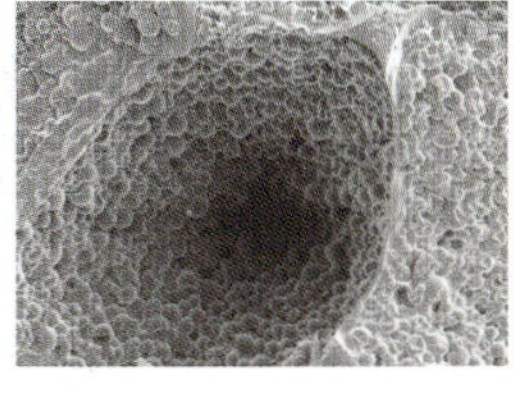

'Thermocules' in foam

▲ **Figure 6-17** The encapsulated phase-changing materials shown are used in outdoor clothing applications. (Courtesy of OUTLAST™)

OTHER

Phase-changing pellets

These products are targeted for architectural applications. The storing and releasing of large amounts of energy via the phase change effect (see Chapter 4) makes their potential use in helping control and maintain thermal environments in a building very attractive. There have been a number of experiments in trying to find an effective way of incorporating phase change capabilities into common building products, including experiments with wallboard products. The problem of containing and distributing the materials has always been problematic. An interesting current approach uses relatively large encapsulated pellets. These pellets can be placed in common floors or walls. They are particularly useful in connection with radiant floor heating approaches. Common techniques of burying hot water pipes into concrete floors can be problematic due to the time-lag problems associated with heat storage and release in concrete. Concrete is slow to heat up and slow to cool down, with the consequence that heat is often being released at the wrong time. Encapsulated phase-changing pellets can respond in a much more timely way and can be incorporated in a light framing system. The positive attributes of radiant floor heating can be achieved without many of the problems associated with heavy mass systems.

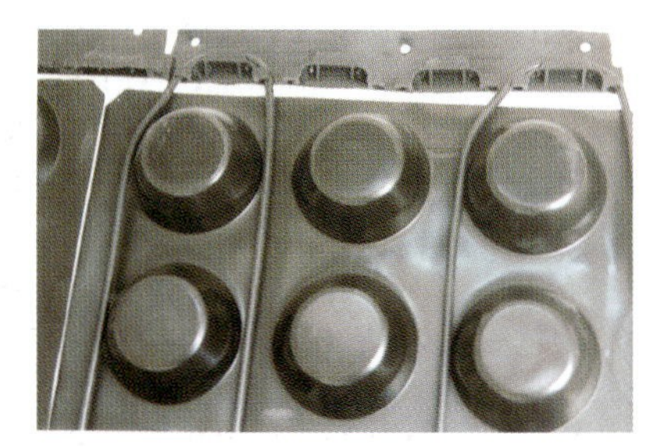

▲ **Figure 6-18** These pellets contain encapsulated phase-changing materials. They are used in radiant heating floor systems. This particular product uses TEAP 29C PCM capsules which are engineered to maintain interior air temperatures at near ideal conditions.

7 Smart components, assemblies and systems

The previous chapter profiled just a few of the many products that are currently using smart materials. We cannot help but be intrigued and fascinated by them. We have found lamps made from crumpled radiant color film, and have tried out the latest thermoelectric mini-refrigerators in our cars. No one would call us Luddites. Nevertheless, when it comes to integrating these technologies into a building, we retreat to convention. We do see a few glowing curtains, and some thermochromic paint on walls, but these tend to be *placed* into the architectural environment, and thus are easily replaced. Serious commitment is required to go any further.

The materials and technologies that are integrated into the building construction, whether it is in the foundation or the electric system, are much more immune to change than the products and ornaments that fill and decorate our buildings. Part of the reason why is because these components and systems must meet fairly rigorous performance requirements, and part is because experiential data is almost non-existent and there is very little information on their longevity. In spite of this disclaimer, however, smart materials have already made many inroads into some of the most prosaic of our building technologies.

The table in Figure 7–1 'maps' smart materials and their relevant property characteristics to current and/or defined architectural applications. With the exception of some of the glazing technologies, most of the current applications tend to be pragmatic and confined to the standard building systems: structural, mechanical and electrical. As these systems are often embedded within the building's infrastructure, many of the smart materials tend to be 'hidden'.

Most noticeable in the mapping is that many smart materials are deployed as sensors. Sensing plays an extremely important although often overshadowed role in the performance of building systems. Even the most routine operation of an HVAC system requires the precise determination of several environmental variables, particularly air temperature and relative humidity. The most visible category for smart material application is in the window and façade systems area, in which these materials are perhaps used as much for their cache as for their performance. It is in this area that architects have become most involved. There are few aspects of a

BUILDING SYSTEM NEEDS	RELEVANT MATERIAL OR SYSTEM CHARACTERISTICS	REPRESENTATIVE APPLICABLE SMART MATERIALS*
Control of solar radiation transmitting through the building envelope	Spectral absorptvity/transmission of envelope materials	Suspended particle panels Liquid crystal panels Photochromics Electrochromics
	Relative position of envelope material	Louver or panel systems - exterior and exterior radiation (light) sensors -- photovoltaics, photoelectrics - controls/actuators -- shape memory alloys, electro- and magnetorestrictive
Control of conductive heat transfer through the building envelope	Thermal conductivity of envelope materials	Thermotropics, phase-change materials
Control of interior heat generation	Heat capacity of interior material	Phase-change materials
	Relative location of heat source	Thermoelectrics
	Lumen/watt energy conversion	Photoluminescents, electroluminescents, light-emitting diodes
Energy delivery	Conversion of ambient energy to electrical energy	Photovoltaics, micro- and meso energy systems (thermoelectrics, fuel cells)
Optimization of lighting systems	Daylight sensing Illuminance measurements Occupancy sensing	Photovoltaics, photoelectrics, pyroelectrics
	Relative size, location and color of source	Light-emitting diodes (LEDs), electroluminescents
Optimization of HVAC systems	Temperature sensing Humidity sensing Occupancy sensing CO_2 and chemical detection	Thermoelectrics, pyroelectrics, biosensors, chemical sensors, optical MEMS
	Relative location of source and/or sink	Thermoelectrics, phase-change materials, heat pipes
Control of structural systems	Stress and deformation monitoring Crack monitoring Stress and deformation control Vibration monitoring and control Euler buckling control	Fiber-optics, piezoelectrics, electrorheologicals (ERs), magnetorheologicals, shape memory alloys

* Many high performance materials (e.g., diochroics, view directional films, and others) may be applicable as well

▲ **Figure 7-1** Mapping of typical building system design needs in relation to potentially applicable smart materials

building that are more important to determining its public presence than the exterior façade. In contrast, lighting systems perhaps have the most impact on the user's perception of the building, and while enormous developments have taken place in this area, they have not percolated as much into the architect's consciousness. Energy systems have steadily become more important as concerns regarding the global environment have mounted. Nevertheless, there remains much confusion as to the role that a building can or should play in the complex web of energy generation and use.

One of the most interesting and least visible of smart material applications in a building involves the monitoring and control of structural systems. Smart materials have a long history in this application, and we are also beginning to look to its substantial use in the civil engineering industry as a model for how we might begin to utilize this technology in our own.

7.1 Façade systems

Façade systems, and particularly glazing, pose an intractable problem for designers. The façade is always bi-directional in that energy transfers in both directions simultaneously. Heat may be conducting to the outside while radiating to the interior, and light entering the building must be balanced with the view to the exterior. The problem of glazing did not emerge until the twentieth century, as it required the development of mechanical HVAC systems to enable the use of lighter weight and transparent façades. At first, the façade systems, albeit lightweight and with an unprecedented amount of glazing, were more opaque than transparent. Constant volume HVAC systems coupled with perimeter systems were more than adequate for mitigating the highly variable thermal loads of the façade, and simple shading devices were used to manage glare. The advent of the energy crisis in the 1970s marked the phasing out of the energy-intensive HVAC systems and their replacement with Variable Air Volume systems.[1] The energy penalty was removed, but at a cost to the thermal stability of the façade which began to loom as a problematic element in the building. Paradoxically, the demise of the CAV system was coupled with a rise in the percentage of glazing on the exterior, further exacerbating the thermal and optical swings of the façade. Compensatory mechanisms and approaches were developed and experimented with, and a host of new technologies were incorporated into the façade or enclosure systems. Glazing was

▲ **Figure 7-2** Dichroic light field from James Carpenter Design Associates. To animate a blank, brick façade, a field of 216 dichroic fins was attached perpendicularly to a large plane of semi-reflective glass

coated with thin films, including low-emissivity, solar reflective, and non-reflective (on the interior faces). Automated louvers were installed in conjunction with energy management control systems to reject excess solar radiation, and elaborate double skin systems, which wrap the building *twice* in glazing, were encouraged for the dampening of the thermal swings. As a result, no other group in the architecture field has embraced smart materials as wholeheartedly as have the designers and engineers responsible for façade and enclosure systems.

Smart materials were envisioned as the ideal technology for providing all of the functions of the super façade, yet would do so simply and seamlessly. Visions of Mike Davies' 'Polyvalent Wall' – a thin skin that combined layers of electrochromics, photovoltaics, conductive glass, thermal radiators, micropore gas-flow sheets and *more* – served as the model of the ultimate façade. In 1984, the seminal theoretician and historian Reyner Banham, while commenting that a 'self-regulating and controllable glass remains little more than a promise', did conclude that if the real energy costs were taken into account, the new technology would prove to be economically viable.[2] His prediction was not far off, as an entire field devoted to the development of smart windows and façades has been premised on their contribution to energy efficiency. Indeed, the lion's share of investment

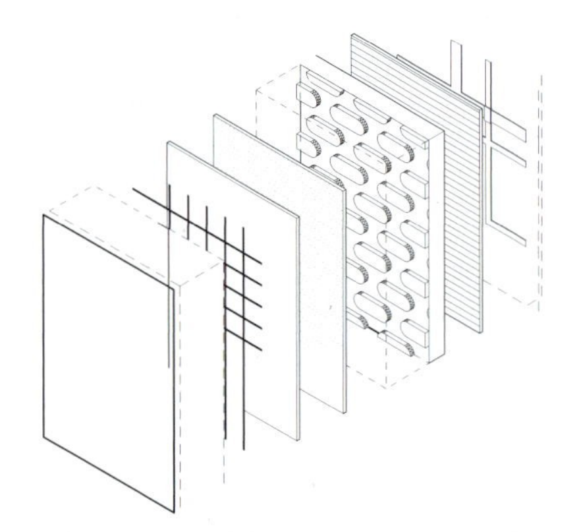

▲ **Figure 7-3** Schematic respresentation of Mike Davies' polyvalent wall. He proposed that the exterior wall could be a thin system with layers of weather skin, sensors and actuators, and photoelectrics

dollars in smart materials for buildings is concentrated on these two systems. Furthermore, in concert with our overview of the contemporary approach to materials in Chapter 1, windows and facades are the signature visual elements of a building, and as such, will *de facto* be of primary interest to architects. As we might expect then, many of the current initiatives taking place in these areas tend toward treating the smart material as a replacement technology that fits within normative design practice.

THE SMART WINDOW

The term 'smart window' has been applied to any system that purports to have an interactive or switchable surface, regardless of whether that surface is a real or virtual window, interior or exterior. For the purposes of this book, we will consider the virtual windows to fall into the category of large panel displays, and concentrate our discussion on exterior glazing and interior partitions.

'Smart' windows will typically possess one or more of the following functions:

- Control of optical transmittance. A shift in the transparency (the optical density) of the material may be used to manage the incident solar radiation, particularly in the visual and near ultraviolet wavelengths. The window would vary from high density (opaque or translucent) for the prevention of direct sun penetration and its associated glare to low density (transparent) as incident light loses intensity.
- Control of thermal transmittance. This is a similar function to that above, but the wavelengths of interest extend into the near infrared region of the spectrum. Heat transmission by radiation can be minimized when appropriate (summer) and maximized for other conditions.
- Control of thermal absorption. Transparency and conductivity tend to correlate with each other, but are relatively independent of the incident radiation. Whenever the inside temperature is higher than the outside temperature, a bi-directional heat flow is established: radiant energy transfers in, while thermal energy transfers out. Altering the absorption of the glazing will ultimately affect the net conductivity, and thus can shift the balance in favor of one or the other direction.
- Control of view. The use of switchable materials to control view is currently the fastest growing application of smart materials in a building. Interior panels and partitions that switch from transparent to translucent allow light to

transmit, but are able to moderate the view by altering the specularity of the material. Exterior store fronts can reveal merchandise in windows selectively, perhaps only when the store is open. A specular material will transmit intact images, whereas a diffuse material will obscure the image.

Depending, then, upon the desired outcome, the designer would choose between several of the different chromogenic materials that were discussed in Chapter 4. While many of the materials can be used interchangeably for the functions – for example electrochromics, liquid crystal and suspended particle will all control optical transmission – each material brings operational and control criteria that can have a significant impact on its *in situ* performance. The most profound difference is between the electrically activated materials versus those that are environmentally activated.

Initially, when architects began to think about smart windows in the late 1980s, their desire was to create a glazing material that responded directly to environmental changes. Photochromic materials had been developed for eyeglasses in which the lens darkened as the incident light increased. This seamlessness in response appealed to building designers, who thought that covering the glazed façades of buildings would provide not only moderation of daylight, but would also help prevent unwanted transmission of solar radiation. Eyeglasses, however, had to address only one condition, that of light incident on the outside of the lens, whereas buildings need to deal with multiple situations, particularly those produced by large swings in exterior temperatures. The most problematic situation is that typical of northern latitudes in the winter: the sun angle is very low, thus producing glare, but exterior temperatures are also low. The ideal responses for the two conditions are the opposite or each other – the sun angle would cause the photochromic to darken reducing the transmitted radiation, but the conductive loss to the exterior would be better offset with a higher rate of transmission. There was also concern about the resulting color of the photochromic in its absorptive state. Depending upon the photosensitive 'doping' chemical added to the glass matrix, the resulting color is either gray or brown – neither of which are particularly desirable for a façade.

Thermochromics are more amenable to the heat issue, but do so by sacrificing control in the visual part of the spectrum. As heat is the activating energy input, thermochromic glazing operates best in the near infrared region of the solar spectrum. The desired switch point is usually set to the interior temperature so that as the temperature of the glazing begins to rise – due either to absorption of solar radiation or to high external temperature – the radiant transmission is reflected rather than transmitted. The application hurdle that thermochromic glazing must overcome is its low transmissivity in the visual part of the spectrum, which currently ranges from about 27 to 35%.[3] Given that the primary reason for a glazed façade is the view, and secondarily, the provision of daylight, thermochromics have been little utilized in the development of smart windows.

Thermotropics respond to the same environmental input as do thermochromics, but the difference in the internal mechanism has given thermotropics broader potential application. Whereas thermochromics switch from transmissive to reflective, thermotropics undergo a change in specularity, resulting in the ability to provide diffuse daylight even as the view is diminished. One feature they offer that is relatively unique is the ability to change the conductivity of the glazing as well as its transmissivity. The phase change that is at the core of any thermotropic results in a substantial reconfiguration in the structure of the material, such that a quite significant change in thermal conductivity could take place. This effect is more pronounced when a hydrogel is used to fill a cavity in double glazing as compared to using a polymer foil as the thermotropic.[4] Some hydrogels can further have two transition states, turning opaque at low as well as high temperatures, rendering them useful for preventing radiant loss from the interior during the winter. Although not nearly as commercially available as the various electrochromic glazing systems, they are expected to become popular for any kind of application, such as skylights, where light rather than view is paramount.

Clearly the major drawback of all three environmentally driven technologies is their inability to 'stop' or 'start' the transition. As discussed earlier, there are numerous circumstances in which the environmental response is not in sync with the interior need. Light, heat and view must cross the glazed façade, and the optimization of a single environmental factor is unlikely to coincide with the desired response to the other environmental conditions. As a result, much more development has been devoted to the various electrically activated chromics, all of which give the user the opportunity to control and balance the often-conflicting behaviors. This control, however, comes with a large penalty. Whereas the environmentally activated technologies can all be incorporated directly into existing façade and window systems, the electrically activated technologies demand a fairly sophisticated support infrastructure. Electrical power must be supplied to each section of glazing, and panel mounting and hardware must be specifically designed and installed to ensure proper operation and protection. Furthermore, to take full advantage of the potential afforded by the ability to turn the system on and off, there is usually an accompanying sensor and logic control system. For example, one popular scenario uses light sensors to optimize the balance between artificial lighting and transmitted daylight. The next generation sensor/control system would take into consideration the heat load of the façade and determine the balance between both types of light with heat, perhaps allowing the artificial lighting to increase if the more economic option is to reduce transmissivity to prevent radiant heat gain. This type of assembly then may push the envelope of our definition of a 'smart material' as the 'intelligence' is fully external, and the actions are not always direct. Nevertheless, electrically activated glazing for building façades has quickly gained commercial viability in just over a decade and remains as the most visible indicator for smart materials in a building.

All three of the electrically activated chromics must have an external logic for their operation, and as a result, the major differences between them are due mostly to the character of the light transmission – whether specular or diffuse, absorbed or reflected. Electrochromics were the first technology that was heavily invested in by glazing and façade manufacturers. As discussed in Chapter 4, the five-layer structure of conductors and electrodes that comprises a typical electrochromic has steadily evolved from an unwieldy system that was easily damaged into a thin coating that can be applied to standard glazing. The reduction in transmissivity is generally proportional across the spectrum such that visual transmissivity drops as much as the infrared transmissivity (each is reduced about 50% between the bleached and the colored states).[5] The need to maximize visual transmissivity while minimizing heat gain has resulted in the development of electrochromics that have high initial intensity in the short wavelength region coupled with low intensity in the long wavelength region. As a result, the colored state of the glazing tends to be blue even though electrochromics can have some spectral variation. Nevertheless, these have become most recommended for building façades due to their ability to maintain spectral transmission, and thus view, from the bleached to the colored states.

Liquid crystal glazing takes advantage of the enormous developments in the liquid crystal arena. As liquid crystals are the primary chromatic technology used in large panel displays, there has already been substantial attention paid to their deployment on large exterior surfaces. As such, unlike the development of electrochromics, which grew exclusively from the desire to use them on building façades, liquid crystal glazing came into the architectural market fully tested and refined. Issues regarding their durability, maintenance, sizing, mounting and packaging (this is in reference to the provision of an electrical supply) had been addressed and at least partially resolved. Architects only had to begin to employ them. In spite of these advantages, however, there are important drawbacks associated with liquid crystal glazing. The first is that when it transforms from its bleached to its colored state, the transmission energy does not change, only its specularity – from specular to diffuse. If we can recall that the primary reason for the chromogenics is to reduce unwanted infrared radiation, then the liquid crystal devices are hardly satisfactory. In addition, unlike the electrochromics, which require power only when the switch in states occurs, liquid crystals require continuous power in their transparent state. And the linear alignment of the crystals when in the transparent stage significantly reduces view from oblique angles. Even with these drawbacks, the use of liquid crystal is rising dramatically for discretionary projects, particularly high end residences and interior partitions where privacy and ample light are more important than energy.

Suspended particle devices are an alternative to liquid crystals for privacy applications, with similar drawbacks. They, too, are not effective for reducing infrared transmission, and they also require continuous power to remain transparent. In addition, they have even less ability for their spectral profile to be tweaked toward one color or another. Their primary advantage over liquid crystals is their ability to permit much more oblique viewing angles.

An issue that arises for all of the electrically activated chromics is the operation of their electrical supply. Unlike the environmentally activated chromics, which may cycle infrequently and further go for long periods without cycling at all, the electrically activated chromics will most likely undergo substantially more frequent switching. Although numerous tests have been mounted to determine the number of cycles before a noticeable degradation in optical properties occurs, there still have not been sufficient field studies to examine cycling in real use. Besides routine operation, the glazing must weather severe environmental conditions and undergo routine maintenance operations like window washing. While one might conclude that the environmentally activated chromics are a safer bet for longevity, we must equally be aware that their chemicals tend to be less stable. Electrical operation is

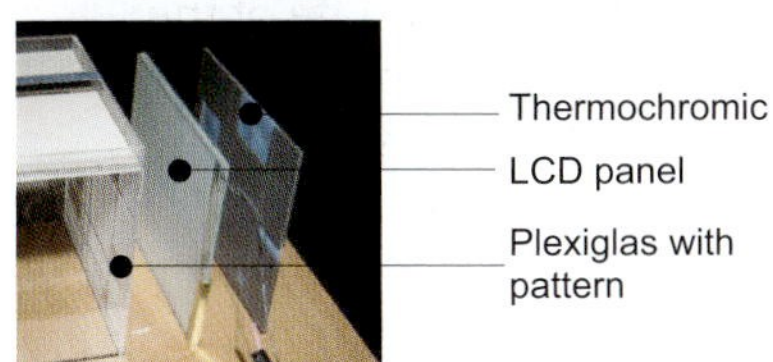

▲ **Figure 7-4** Design experiment: the patterns in this wall study vary with changing temperature and with the on–off state of the LCD panel. (Yun Hsueh)

also important insofar as we consider when voltage or current must be supplied. Because electrochromics only require power to switch from one state to another, and no power to remain at either state, they can be supplied with batteries. Liquid crystals and suspended particles need continuous power to stay transparent, and as a result, require an electrical infrastructure to supply the façade. The continuous power also negates any energy savings they might produce.

The table in Figure 7–5 summarizes the salient design features of the various chromogenics. The first question that must be asked is what result we want in the interior. Do we wish to reduce the infrared radiation transmitting through the glazing but not lose the view? Are we willing to lose the view, but not the light? Is control of glare important? In the table, view is determined by specularity – specular transmission provides view, whereas diffuse transmission produces an opaque surface. A glazing that has specular to specular transmission will not impact the view, but will reduce the intensity of the transmitted radiation. Different types of coatings will determine in which bandwidth that reduction will primarily take place. Obviously, for control of heat, the ideal glazing material would be little impacted in the visual range, but show a markedly reduced transmission in the

SMART WINDOWS				
System type	Spectral response (bleached to colored)	Interior result visual	Interior result thermal	Input energy
Photochromic	Specular to specular transmission at high UV levels	Reduction in intensity but still transparent	Reduction in transmitted radiation	UV radiation
Thermochromic	Specular to specular transmission at high IR levels	Reduction in intensity but still transparent	Reduction in transmitted radiation	Heat (high surface temperature)
Thermotropic	Specular to diffuse transmission at high and low temperatures	Reduction in intensity and visibility, becomes diffuse	Reduction in transmitted radiation, emitted radiation, and conductivity	Heat (high and/or low surface temperature)
Electrochromic*	Specular to specular transmission toward short wavelength region (blue)	Reduction in intensity	Proportional reduction in transmitted radiation	Voltage or current pulse
Liquid crystal*	Specular to diffuse transmission	Minimal reduction in intensity, reduction in visibility, becomes diffuse	Minimal impact on transmitted radiation	Voltage
Suspended particle*	Specular to diffuse transmission	Reduction in intensity and visibility, becomes diffuse	Minimal impact on transmitted radiation	Current

* indicates that a control system and associated electrical supply are required

▲ **Figure 7-5** Comparison of smart window features

infrared region. On the other hand, for glare control, a reduction in the intensity of the visual transmission is important. If the desire is for privacy while maximizing the available daylight, then liquid crystals are the best option. If the need is to minimize heat exchange through the material, then a thermotropic is the best option.

7.2 Lighting systems

The production of artificial (electrical) light is the most inefficient process in a building. As such, there has been a concerted effort to improve the efficiency of the individual lamps. Fluorescents are up to five times more efficient than incandescents, and high intensity discharge (HID) lamps are twice as efficient as fluorescents. But, as discussed in Chapter 3, the production of light from electricity is what is known as an uphill energy conversion, and thus the theoretical efficiency is extremely low. The efforts devoted to improving lamp efficiency are netting smaller and smaller energy savings as the theoretical limit is being approached. Smart materials can have a major impact on energy use, even insofar as they are not that much more efficient at producing light than are conventional systems. The fundamental savings will come from the lighting *systems* that smart materials enable, rather than from any single illumination source.

The current approach to lighting was developed nearly a century ago, and like HVAC systems has seen very little change.[6] Ambient lighting, or space lighting, emerged as the focus of lighting design, and it has remained as that focus, even as we have learned much more about not only the behavior of light, but also the processes of the human visual system. Without repeating the information presented in Chapter 3 regarding the human eye and light, we do need to recall that the eye responds only to difference and not to constancy. Ambient light privileges constancy, and as perhaps an enigmatic result, the more ambient light that is provided, the more task light someone will need in order to see.

Although the understanding that contrast in light levels is more important than the level itself is now becoming more widespread, existing lighting technology remains geared toward ambient light. The beam spread of fluorescents demands a regular pattern of fixtures, and the intensity level of HID lamps requires a mounting height far above eye level. In the late 19th century, as artificial lighting began to enter the marketplace, incandescents were described as being able to 'divide' light. This idea of division was in stark contrast to the dominating light produced by the preeminent arc lamp, the intensity of which was so high that entire streets could purportedly be illuminated with a single lamp. A century later, we return to this idea of division, looking to smart materials to enable a discretely designed lighting system that allows for direct control of light to the eye, rather than light to the building.

FIBER-OPTIC SYSTEMS

We start with fiber-optics even though they are not technically smart materials; no transformation takes place in a fiber-optic, it is only a conduit for light. The use of fiber-optics for illumination, however, demands a radical shift in the way one thinks about lighting. Each optical cable will emit a fraction of the light emitted from a more typical lamp, but the light can be more productive. Ambient lighting systems fall prey to inverse square losses, the intensity drops off with the square of the distance. The light-emitting end of the fiber-optic can be placed almost anywhere, and thus can be quite close to the object or surface being illuminated. The tiny amount of light emitted may deliver the same lumens to the desired location as light being emitted from a ceiling fixture at more than an order of magnitude greater intensity. Contrast can also be locally and directly controlled. As we can see, then, fiber-optic lighting possesses two of the important characteristics of smart materials – they are direct and selective.

Fiber-optic lighting offers other advantages over conventional systems. The source of light is remote in comparison to where it is delivered. As a result, the heat from the source is also remote. Lighting, as an inefficient process, produces more heat than light such that about one-third of a building's air-conditioning load is simply to remove the excess heat generated by the lamps. Not only does a remote source save energy, but it protects the lighted objects from heat damage and possibly even fire. Since no electrical or mechanical components are required beyond those at the source location, electrical infrastructure can be reduced and maintenance is simplified. Color control and UV/IR filtering can easily be incorporated, expanding the versatility not only of the system but of each individual cable. These advantages, particularly in regard to the heat reduction and UV control, have rendered fiber-optics the choice for museum exhibit lighting and for display case illumination. The majority of other architectural uses, however, tend to be decorative, utilizing the point of light at the emitting end of the cable as a feature rather than for illumination. Even though there are good models for the effective and efficient use of fiber-optic illumination, the

Lighting

Typical illuminator

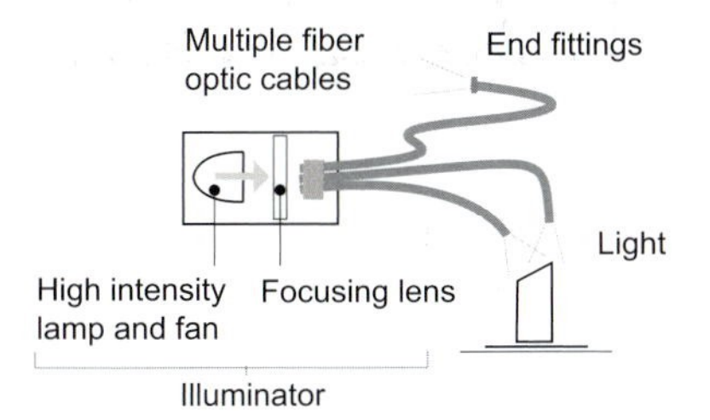

▲ **Figure 7-6** Fiber-optic lighting. Multiple cables can be served from a single lamp. The lamp heat and fan noise is removed from the object being illuminated

paradigm of the ambiently lit interior is so pervasive that only those applications with critical requirements have utilized this discrete approach to lighting.

A fiber-optic lighting systems is comprised of three major components:

- *Illuminator:* this houses the light supply for the fiber-optics. The source of light can be anything, from LEDs to halogen, metal halide, or even solar radiation. Key features of the source are its color and intensity; the greater the intensity, the greater the number of emitting ends, called tails, that are possible. Greater intensity also enables longer length of the tails, up to 75 feet. The light source generates a large amount of heat which then must be dissipated by heat sinks and/or fans. Reflectors and lenses will narrow the light beam as much as possible to fit within the cone of acceptance (this is determined from the critical angle of the strand medium). Light must enter the acceptance cone, so the more collimated the source, the more efficient the transformation will be. Color wheels and other filters are often included in the illuminator to create special lighting effects or eliminate unwanted UV. Electronic controls, including ballasts and dimmers, are also housed in the illuminator.
- *Cable or harness:* fiber-optics for lighting are either solid core or stranded fiber, both of which are bundled into cable form and sheathed with a protective covering. (No cladding is used.) The emitting end will most likely be split into multiple tails, each one providing distinct illumination, while the source end will be bound as a single cable and connected into a coupler, which is then connected to the illuminator. The entire cable assembly, including the coupler, is referred to as a harness.
- *End fittings:* for end emitters, the tail ends will need to be secured or mounted in some manner, and the primary purpose of the end fittings, which are usually threaded, is to allow this. The fittings can also house individual lenses and filters so that the light emitted from each tail end can be controlled separately.

Unlike the fiber-optics used for data transmission, imaging and sensing, those used for lighting are coarser and do not have the same rigorous requirements regarding optical defects. The most common material for the strand is plastic rather than glass. Plastic, usually polymethyl methacrylate (PMMA), is less efficient at the source, with an acceptance angle of only 35°.[7] It also brings a limitation on the bending radius, which is generally recommended to be no smaller than 5 to 10 times the cable diameter. Nevertheless, in the visual part of the spectrum, plastic exhibits similar transmission characteristics as glass, and is further much more flexible to install. It can be cut in the field, the ends can be finished in a variety of ways, and it can be used for side emission as well as end emission (side lighting systems use a clear PVC sheathing).

The more impurities in the fiber, the more the attenuation. PMMA strands lose about 2% intensity per foot depending upon the strand size, with smaller strands losing less. The length of plastic is therefore limited, with lengths of 30 ft considered to be the maximum for end-emitting and 5 ft for side-emitting. Attenuation is also wavelength-dependent, so the longer the cable, the more green or yellow the light becomes.

Side-emitting and discrete-emitting fiber-optics have opened up many new possibilities for uses in buildings. Selective etching along the strand length alters the surface angle enough so that certain angles will no longer internally reflect, but emit along the fiber. The fiber is then a 'light rope' and this technology has quickly overtaken both neon and cold cathode lighting for decorative uses and signage. The fiber-optic 'rope' brings several advantages over its competition; it is bendable, dimmable and amenable to many types of color and optical effects.

Fiber-optics are also an ideal companion for solar-based lighting. Heliostats and collectors can be positioned remotely, so as to take best advantage of the available daylight, and when coupled with a lens system, most likely Fresnel, the light can be concentrated and directed into the harness. Areas that had no possibility of utilizing natural light can bring in full spectrum light that maintains a connection to the transiency of the outdoors.

SOLID STATE

Solid state lighting is a large category that refers to any type of device that uses semi-conducting materials to convert electricity into light. Essentially the same principle that drives a photovoltaic, but operated in reverse, the solid state mechanism represents the first major introduction of a new mode of light generation since the introduction of fluorescents at the 1939 World's Fair. In this category can be found some of the most innovative new smart technologies, including organic light-emitting diodes (OLEDs) and light-emitting polymers (OLPs), but the workhorse technology, and by far the largest occupant of this group, are inorganic light-emitting diodes

(LEDs). The use of LEDs for task lighting, signage, outdoor lighting, façade illumination, traffic signals, mood lighting, large panel displays and other applications is a far cry from the 1980s when LEDs were primarily used as indicator lights, letting us know that our oven was on, or that our car alarm had been activated.

We might consider LEDs as the 'smart' version of fiber-optics, as, in addition to being discrete and direct, they are also self-actuating, immediate and transient. Furthermore, while fiber-optics allow for the division of light, LEDs allow for its recombination in arrays of any multiple. We could almost consider fiber-optics as an intermediate placeholder for the spot that will eventually be taken over by LEDs. The advantages of LEDs over any other commercially available lighting system are profound. Besides their small dimensions which allow their deployment in spaces unable to be illuminated with any other means (fiber-optics still must be tethered to a rather large illuminator), the spectral qualities of the light can be precisely controlled, eliminating both the infrared radiation that accompanies incandescents and the ultraviolet radiation that is associated with most discharge lamps. Beam spreads can be controlled or concentrated at the source, reducing the need for elaborate luminaires with large filters, reflectors and lenses. Considered as their largest drawback is their low efficacy, which at about 20–30 lumens/watt still beats out the typical incandescent.

The companies that produce LEDs have been feverishly working toward ever-higher efficacies, with 100 lumens/watt considered as a necessary goal in order to compete in the fluorescent market. In preparation for that competition, some manufacturers have already begun to produce arrays in strips that can be packaged into long tubes. As such, much of the experimentation and development is focused on ambient lighting. This is very much a 'chicken and egg' dilemma: ambient lighting emerged in the 1940s as a strategy to develop a market for fluorescents, so is it the existing technology that is controlling the manner in which new technologies are developed rather than an environmental or human consideration?

Excepting, of course, signage and decorative uses, lighting in buildings has primarily been for the illumination of space and objects. Incandescents generally serve as the object illuminators, whereas the discharge lamps are intended for space lighting. Fiber-optics have paved the way for object lighting with discrete sources, but LEDs are not routinely seen in this application. This may be due to the ease of installation and low cost of incandescents, or it may simply be due to the

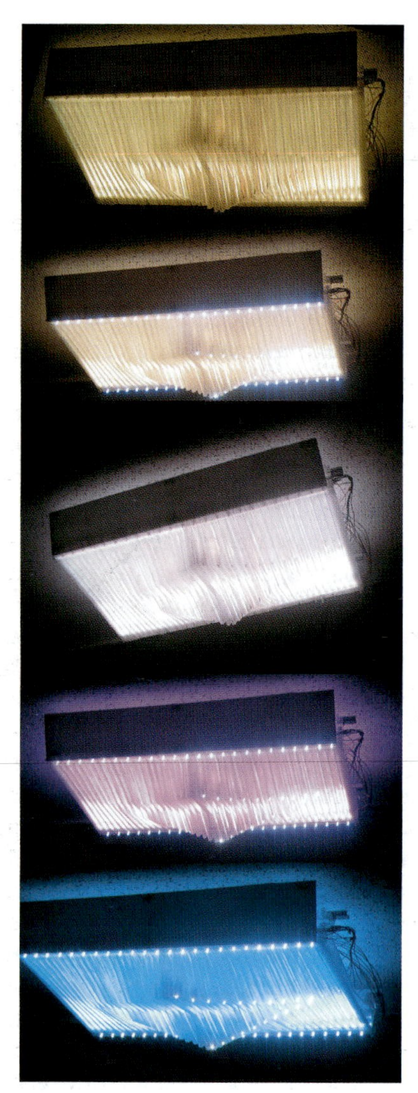

▲ **Figure 7-7** Experiments in architectural LED lighting by Maria Thompson and Rita Saad of MIT in cooperation with OSRAM Opto Semiconductor group. The tiles represent experimentation with LED color, angle and beam spread in combination with the refractive properties of a medium

'newness' of LEDs and not many people, other than lighting designers, are yet familiar with the specifics of their use and their wide-ranging attributes.

Clearly, if the efforts in place at the major manufacturers succeed in their goals of higher efficacy, higher flux and lower cost, then LEDs will be excellent replacements for both incandescents and fluorescents. But rather than trying to make this new technology behave the same way as the existing, can we not begin to explore how it might be different? For example, how best might we take advantage of dynamic control as there is no other lighting technology that offers both dynamism and transiency? The transiency is unparalleled – any color at any intensity at any time, at the size of a pixel to that of a large surface.

This is an area in which architects and designers could play an important role in opening up new possibilities. We have seen some initial forays into exploring the color possibilities of LEDs in several interior 'mood' installations – for example, a room where occupants can dial in their favorite color, and a hotel dining room where the colors match the cuisine. LED lighting is increasingly becoming popular for bars looking to differentiate their ambience from their competitors, incorporating subtly moving patterns as well as color transformations. LEDs are particularly suited for pattern making and color variability, since both effects can be produced by incandescent systems with color wheels.

It is at this point that we should return to our earlier discussion in Chapter 3 regarding the physics of light and the environment of the eye. Our current lighting systems neither exploit the fundamental characteristics of light nor do they address the particularities of the eye's stimulus and response behavior. If light obeys the laws of geometric optics, why are we not using the principles of refraction and reflection to precisely manipulate the path of light? If color belongs to light, and surfaces are only capable of subtracting color, then how else should we be thinking about color other than in a demonstrative way? LEDs offer us the possibility to design our luminous environment directly, rather than indirectly through ambient lighting. Furthermore, since our eye only responds to difference – in color and in luminance – then the opportunity exists to discretely produce small step changes that can have a profound impact on our perception of our surroundings. LEDs allow for a luminous articulation that is not possible with any other lighting system.

Although inorganic LEDs have drawn the majority of efforts in the development of new solid state lighting technologies, there is considerable speculation regarding the entry of organic materials. Organic LEDs or OLED can be split into two types: SMOLED (Small Molecular OLED) and PLED, or (Polymer OLED, also referred to as POLED). The great advantage of the organic types is that they are based on thin film technology. Conventional LED systems are inherently constructed from point sources, and color change requires the right combination of the individual diodes in a particular region that then must be covered with a diffusing layer for homogeneity. The thin films, in contrast, can provide any type of color and light distribution, regardless of how the individual pixels are arrayed. And since each pixel is fully addressable, OLEDs can be used as displays, television screens and anything else that might require dynamic spectral and luminance control. Perhaps the most provocative aspect of OLEDs is that the films are flexible, able to be molded around curves, and transparent, which allows almost any substrate.

The progressive miniaturization and discretization of lighting technologies is beginning to introduce a new paradigm in lighting design. Regardless of the technology or system used, including that of daylight, traditional lighting systems are designed with respect to the delivery of light to designated surfaces. Current standards in lighting require the provision of a specified illuminance to task surfaces and floor areas. The primary interests in the source are its color and efficacy. Because sources are generally some distance away from the surface to be illuminated, their lumen output is often one to two orders of magnitude higher than what can be accommodated for in the field of vision. A wide array of shielding devices – baffles, diffusers, troffers, coves – are utilized to reduce the visibility of the light source, thus preventing glare.

Our new found ability to 'look' at light sources requires a radical reconsideration of the deployment of light sources. Rather than delivering illumination to a surface, they can produce a luminous surface. The source can be integral, and the ability to place it anywhere opens up other uses for the light, particularly the delivery of information. In the lobby project for the Bear Stearns Company in New York, James Carpenter Design Associates used two different advanced lighting technologies in conjunction with high performance optical materials directly to place the light sources within the field of view. Glass vitrines flank the security desk, framing it in a soft, but subtly shifting gold light. Several layers of different glasses, including dichroics, add to the indeterminate nature of these vitrines, and four columns of end-emitting fiber-optics provide the illumination – both indirectly as the light optically transverses the glasses, and directly as discrete moments of sparkle. On the same wall, freestanding ticker

▲ **Figure 7-8** Fiber-optics, dichroic glasses and LEDs were used by James Carpenter Design Associates in this lobby installation for Bear Stearns in New York. The green zone is produced with fiber-optics and dichroic glass, it serves as a soft contrast to the moving blue LED information screens. (James Carpenter)

tape columns are configured in such a way that at one position the blue LED display communicates live stockmarket information, and at other positions etched glass surfaces create a blurred movement that complements and contrasts with the gentler vitrines.

7.3 Energy systems

There are three types of energy needs in a building: thermal, mechanical and electrical. Thermal energy is necessary for heating and cooling of spaces, refrigeration, water heating and cooking. Mechanical energy is necessary for fans, motors, compressors, pumps and many appliances. Electrical energy is only directly required for lighting and peripheral equipment such as televisions and computers. These may be the needs, but the sources for supplying those needs are a different energy type altogether. There is no reasonable manner for supplying mechanical energy directly in a building, and it is not possible for any heat-rejecting need – refrigeration, space cooling – to operate without a compressor. Electrical energy, which is a relatively minor need, becomes one of the major energy supplies in a building, as it is the only source that can power mechanical equipment. Indeed, electrical energy could, and often does, take over the remaining thermal needs of heating, cooking and water heating. As a result, two-thirds of a building's energy use is due to electricity, and, perhaps even more disturbing, two-thirds of the electricity use in the United States is due to buildings. Reducing the electricity used in a building looms as one of the key targets for reducing worldwide greenhouse gas emissions.

Clearly, then, developing and investing in systems that would reduce electricity use would seem to make sense, and furthermore, the unique energy transferring characteristics of many smart materials should render them ideal for building uses. This is an area, however, that has not received as much investigation as have the various lighting and façade systems. Most of the attention has been devoted to replacing part of the fossil-fuel-based electricity generation with photovoltaic generation rather than investigating new approaches for reducing energy consumption. While one might assume that the fuel mix for electricity generation is not a building issue, as it is the regional utility that is most affected, the unit size of photovoltaics is small enough that they can easily be scaled to building size. Obviously, if the photovoltaic is connected to the electrical grid, then its specific location in the grid is not relevant. Nevertheless, photovoltaics have emerged as the front-line strategy for 'green' buildings. Other systems that use smart materials, particularly thermoelectrics, have received relatively little investigation as compared to that devoted to building-sized photovoltaics.

PHOTOVOLTAICS

Photovoltaics were essentially the provenance of NASA until about two decades ago when large-scale photovoltaic generating facilities were first built. The high cost and low efficiency of these facilities prevented their widespread adoption as it was more effective to use the solar energy to produce steam in an intermediate step rather than to produce electricity directly. Concurrently, PVs as replacements for batteries began emerging in small products such as calculators and watches. Although PVs were commonly used in remote areas to power villages, houses and even offshore platforms without an electrical supply, they did not gain favor in grid-connected buildings until the electrical industry was deregulated. Through its 'Million Solar Roofs' initiative launched in 1997, the US Department of Energy has been encouraging, and rewarding (through tax credits), the widespread privatization of grid-connected PVs. Relieving the utilities of the high investment burden of moving to carbon-free generation, the push for building-scale distributed PVs was also intended to speed up the transition to solar

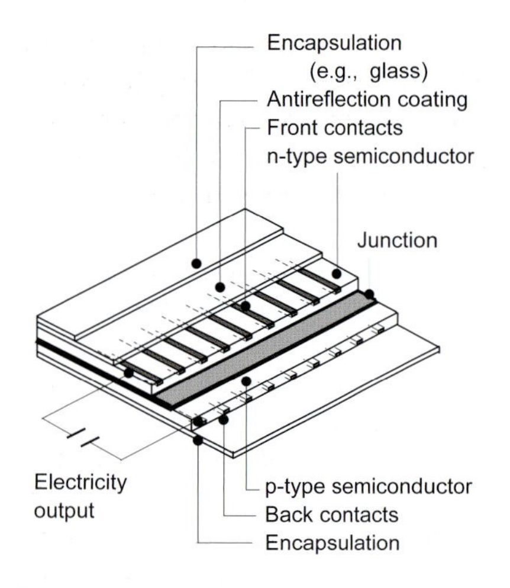

▲ **Figure 7-9** Schematic layout of a photovoltaic cell

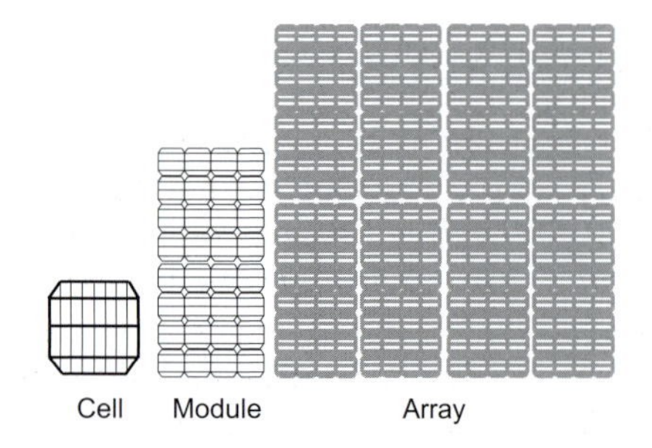

▲ **Figure 7-10** Components of a PV system

technologies. The term 'building-integrated photovoltaics (BIPV)' is now a part of every architect's vocabulary.

Some basic understanding about PV operation is needed before the decision is made to install a system. A typical photovoltaic cell produces about 2 watts. The cells are connected in series to form modules, and the modules are connected in parallel to form arrays. Series connection is necessary to build up to an adequate operational voltage, but it is then vulnerable to any weak link in the connection. If individual cells in the modules are unevenly illuminated, perhaps due to shading from trees or dirt accumulation, the unlit cells will dissipate the power produced by the illuminated cells. Night-time causes a problem as well, blocking diodes are necessary to prevent PV systems from *drawing* electricity when they are not illuminated, but not without a penalty. The penalty is an output power loss. As a result, the efficiency of the module is approximately 20–25% (this is before one even takes the PV cell efficiency into account). Thin film technologies are not subject to inter-cell losses as they are manufactured directly as modules, but they still suffer the inverter and transformer losses.

Stand-alone systems store the generated power in batteries, whereas the intention of BIPVs is to interconnect with the utility grid. The appropriate matching of PV operation with the utility is known as 'balance of system'. This is not a small issue, since both are dynamic systems. There is power-conditioning equipment that manages this transfer, but it too reduces overall efficiency. Just the inverters for converting the DC power of the array to AC power for the utility operate at efficiencies between 70 and 90%. (Stand-alone systems also require 'balance of system': batteries need charge controllers to prevent overcharging and excessive discharging.) Each additional piece of equipment needed in the system reduces the overall efficiency.

The efficiency of the individual solar cell by itself has received much of the attention in research and development. While solar cell efficiencies are continuing to increase, and are now about 8% for thin film to about 18% for single crystal silicon, we must recall that the conversion of radiation to electricity is an uphill process, and thus one in which the theoretical efficiency will be limited.[8] While solar energy is certainly copious, the low efficiency would not seem to be an issue with the exception of cost, but the principle of exergy tells us that energy lost due to efficiency is converted to heat. The efficiency calculation for PVs takes into consideration solar radiation that is reflected, so not all of the efficiency drop creates heat; nevertheless, 40–45% of incident solar radiation on a module does produce heat. PVs are sensitive to heat such that as the cell temperature increases, the efficiency begins to drop, which in turn further increases its temperature. The silicon-based cells are most sensitive to heat; their maximum efficiency occurs at 0 °C and it drops to half as the cell reaches room temperature. Standoff or rack mounting allows for passive cooling of the arrays, as all sides are exposed to air movement. Studies have shown that PVs directly mounted on building surfaces operate at 18 °C higher than those that are standoff mounted.[9] Thin film PVs are less sensitive to heat, but they are the least efficient to begin with.

Regardless of the type of PV, from a single crystal to the most advanced thin film, all share concerns that include soiling and orientation. Estimated losses due to soiling vary from 5% to 10% a year. The optimum tilt angle, to minimize cosine law reductions in intensity, is 90% of the latitude of the site, and the optimum azimuth varies from due south to west depending on whether the generated power is needed for peak load production. In Houston, the optimal tilt of an array would then be 27°, in Boston, 38° and in London, it would be 47°. Given that building façades are typically at 90°, PVs installed anywhere other than on a tilted roof rack facing south will have diminished performance.

Utility integrated PVs thus pose a dilemma for architects. Once connected to the grid, a PV installation has no direct relationship with the energy that the building is consuming. Yet, building installations are perhaps the most practical method for distributing the investment cost and for encouraging more rapid adoption than is likely to occur at the utility scale. Rather than attempting to make photovoltaics fit the building, i.e. through thin film PVs on glazing surfaces or PVs as built into roof shingles, we should perhaps ask how building installations could most effectively contribute to optimal PV performance. This requires that we collaborate with electrical engineers and power engineers to develop better solutions.

MICRO- AND MESO-ENERGY SYSTEMS

Photovoltaics was the first semiconductor thermal technology to make its way into buildings, and there are currently several more coming down the pike. These newer systems, however, are intended to have a direct relationship with the building's thermal system, i.e. heat or cool the building, rather than supply the utility. The efforts in this arena are a direct offshoot of unrelated research taking place in areas as diverse as electronics cooling and miniature battery development.

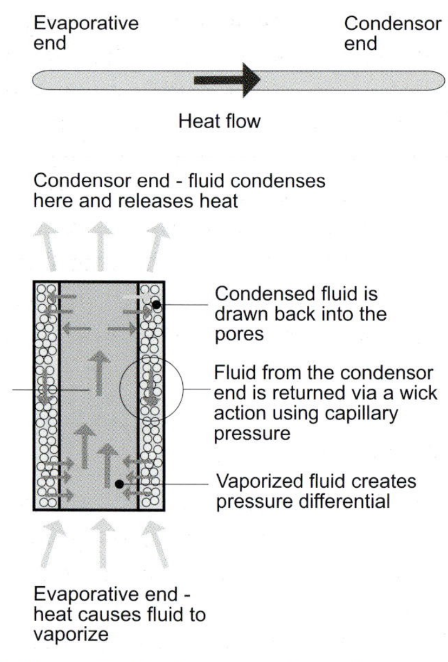

▲ **Figure 7-11** Heat pipe: this self-contained device is extremely efficient in transferring heat from one location to another

As discussed in Chapter 5, the development of MEMs technology has propelled many scientists and engineers to re-examine phenomena at much smaller scales. Light, as a micro-scale behavior, can be as effectively acted on at its own scale as it can at much larger scales, and this has led to a surge in optical MEMs that use tiny mirrors to produce quite large effects. The same is true in thermal behavior. It was not until the electronics industry started applying micro-cooling for chip thermal management that processor speed was able to escape what had been seen as a thermal limit (thus allowing Moore's Law to stay on track for another few years!). Thermoelectrics, or Peltier devices, became the heat sink for the chip, and when combined with heat spreaders and heat pipes, were able quickly to move heat away, particularly in laptops, where size prevented the use of a fan. It was not long before researchers began to speculate on whether these tiny heat pumps could be used for building applications as well.

The initial research was directed toward the replacement of large HVAC systems with micro- and meso-scale technologies. Unfortunately, efforts were slowed for many years as the approach privileged the HVAC system in that the micro- and meso-scale devices were expected to produce the same behavior. One group of researchers simply ganged thermoelectrics together in series to produce a closet-sized system that could be directly substituted for a building heat pump. While the system worked, its efficiency was quite low, as thermoelectrics, regardless of their remarkable qualities, are still uphill processes, thermodynamically speaking. Another group of researchers went in the other direction, repackaging the meso-scale thermoelectrics as a micro-scale film. Their concept, an interesting one, was to produce a sheet of heat pumps that could be applied in rooms just like wallpaper. Nevertheless, they were still thinking subordinate to the existing technology, and abandoned the project when the sheet was unable to match the behavior of a conventional HVAC system.

Micro- and meso-scale research of thermal technologies has since retrenched, and currently efforts are being focused on better scale matching of devices and behaviors. Researchers have recognized that the HVAC system is an anomaly, and these small devices could be more effectively used if deployed directly on the object that needs the thermal management, rather than indirectly through the surrounding environment. While still not available for building uses, these technologies will soon profoundly affect the constituency of our thermal environment. Soon to be available, however, are micro-heaters, although their current size, about that of the

▲ **Figure 7-12** This small methanol fuel cell is an electrochemical device that takes a methanol water solution and uses it to make electricity. It can be recharged instantly by adding the liquid fuel, which is easier to handle than the gases used by other fuel cells. (NASA)

palm of one's hand, puts them more in the meso-range. These devices are ideal for water heating, and other direct uses such as baseboard heating. Meso fuel cells are being developed, and they will have many potential uses, particularly the powering of larger devices. Point loads, such as lighting and computers, could be controlled directly, insulation could be turned on and off by altering its thermal profile, and contaminated air could be corralled and separated from the occupants. The enormity of the possible change requires that *both* engineers and architects be willing to let go of the constraints of conventional practice.

7.4 Structural systems

Structures behave in complex ways when subjected to forces that originate externally to the structure (winds, earthquakes) or are due to its use context, or even its own dead weight. Specific members, for example, bend or tend to buckle. Connections tend to shear apart. Whole structures move about in different modes during dynamically acting earthquake or wind loadings. Designers have been studying these phenomena for hundreds of years, and by now the structural design profession is quite a sophisticated one. The field is strengthened by many computer-based structural analysis simulation programs that help a designer predict the behavior of even the highly complex structures under loadings. Nonetheless, there are still many problematic aspects to designing safe and efficient structures for necessary strength and stiffness criteria. For example, providing safe structural responses to wind and earthquake loadings that produce complex dynamic effects in structures remains challenging. The field of smart materials has opened up some new avenues for solving some old problems in structures, while at the same time opening up some new design possibilities that extend to new areas called 'smart structures'.

One of the fascinations that the field of structures has for many designers is that structural configurations seem to acquire an active life when subjected to forces associated with external loadings (wind, earthquakes), their use context (occupancy loads) or their own self-weights. They change shapes, they move back and forth. They tend to bend, and sometimes break. The by now tired analogy so often cited of how a building or bridge structure works by comparing it with that of the skeleton of human body still serves a useful purpose when considering what a 'smart structure' might be. In the analogy, however, there is a suggestion that the skeleton alone provides the human with its structural capabilities, which is obviously wrong. There is a whole system of other sensing and actuating elements that are interconnected with the skeleton, e.g., tendons and muscles. Without these interconnected elements, the skeleton would collapse. The overall system is not a static one – there are many active actions – e.g., muscles that contract and exert forces at critical locations – that provide responses to even the simplest changing external condition. Consider the complexity of the human body's response when a human reaches out to pick up even a small weight at arm's length. Not only do the arm structures spring into action, but literally the whole body does as well, including stance changes. Many of these actions can occur literally instinctively and exhibit what might easily be called 'smart behaviors' and are associated with what are now generally called 'smart structures'. There are also other responses and more general strategies, however, that might be dictated by the human's intelligence. Knowing that an unusual operation is in the offing, for example, a human might choose to reposition his or her stance *a priori* in order to better accommodate the impending condition. In these cognition-controlled processes, we have a suggestion of what might be called an 'intelligent structure' of the kind described in the following chapter.

In this section we begin considering what is realistically meant at the present time by the term 'smart structure'. We begin with a series of techniques, now generally called 'structural health monitoring', which serve continuously to monitor a structure for any damage that may be present. The idea is an attractive one. In much the same way that our body detects a problem with a sprain, a broken bone, or a cut, it is easy to imagine structures with capabilities for monitoring their own health and providing various alerts to users as a prelude to suitable responses. One of the exciting new developments in this general area is the development of what might be called 'smart skins'. These are surface structures that have sensing capabilities throughout their extents. For example, there have been experiments with micro-sized piezoelectric particles distributed throughout a surface that detect when abrasions occur. Using the very stuff of the material itself to provide a sensing and reporting function is suggestive of what a 'smart structural material' might ultimately become.

Following the section on structural health monitoring, ways of controlling vibrations and other phenomena of concern in a structural design context are explored. In general, most current approaches to smart structures have as their objective a capability for sensing an outside effect and providing a suitable structural response to it. In the following, both passive and active smart structures are reviewed. A passive system normally has the objective of minimizing the effects of an unwanted phenomenon through the simplest responsive means possible. An active system generally means that a means of controlling an unwanted phenomenon is provided via force applications or other techniques. Current approaches typically consist of the structure, a sensor system, an actuator system and a control system. The control system includes a microprocessor that analyzes input data, relates it to a mathematical model of the structural behavior of interest, and sends out suitable output signals to the actuator systems that provide requisite balancing or responding forces. The continued development of smart materials may ultimately allow many of these functions to be integral to the structure rather than exist as separate components.

STRUCTURAL HEALTH MONITORING

The term 'structural health' has recently gained currency as a way of describing a broad field of research and development work that focuses on understanding damage or deterioration in structural systems. The sources of damage are many and may be caused by a variety of external environmental forces (earthquakes, winds, extreme temperatures) and impacts, overloads, or vibrations associated with the use context. Damages may also ensue from initial manufacturing or construction defects, including material and fabrication problems, misalignments, inadequate connections and so forth. Structural performance may also deteriorate due to time or environmentally induced changes in a material's mechanical properties (the modulus of elasticity of some materials, for example, can lower when the material is in a wet environment).

Over the years a great many inspection techniques have been developed for detection, assessment and monitoring of damage and deterioration of structures, including various non-destructive evaluation techniques that may be on-site or remotely utilized. These range from various visually based inspection technologies to the ubiquitous strain gage that is directly affixed to a member. The field has a longstanding history and is huge. The recent surge of developments in the smart materials area, however, has significantly added new capabilities. Also, developments in aerospace applications and other high-end industries have led to interesting design approaches that directly incorporate sensor-based detection and analysis systems that continuously monitor structural health in sophisticated ways that may provide useful models for other application domains.

According to the Japanese researchers Fukuda and Kosaka from Osaka City University, four major new damage assessment approaches are particularly interesting. These are based on fiber-optics, piezoelectrics, magnetostrictives and electric resistance technologies.[10]

Embedded fiber-optic cables can be used to assess breaks, sharp bends, vibrations, strains (deformations) and other occurrences in the base material. Simple breaks and bends can be assessed quite easily. Strain and vibration measurement is more difficult. Depending on what is to be measured, applicable technologies may or may not be computationally assisted. Most have a troublesome sensitivity to temperature conditions. All are based on some type of analysis of the characteristics of the light transmitted through the embedded fiber-optic cable. Deformations, fractures, bends or other effects associated with actual or impending damage to the base material change or affect the characteristics of the transmitted light in some way. Detecting and interpreting the meaning of these changes requires considerable technical understanding of fiber-optic technologies that will only be touched on here.

Following Fukuda and Kosaka*, fiber-optic technologies* for damage assessment include intensity-based sensors, interferometric sensors, polarimetric sensors, Raman scattering sensors, brillouin scattering sensors and several related specialized technologies (EFPI or FBG sensors). Intensity-based approaches measure the intensity of transmitted light, typically generated from a LED source and measured on the other end via a photodetector. These are relatively inexpensive technologies. Breaks or fractures can be detected when transmission levels drop. Interferometric sensors are more complex. The EFPI sensor, for example, is an insert tube that consists of two small half mirrors at the ends of adjoining fiber-optic cables. The device can measure light interference patterns that can be subsequently interpreted to determine strain levels. It has a short sensing length. The FBG (fiber Bragg grating), by contrast, measures wavelength shifts caused by strain variations along its length. Other techniques, such as using brillouin scattering, depend on analyzing frequency peak shifts of the waveforms transmitted along a length of fiber-optic cable. This technique can measure both strain and temperature. In all of these examples, data-gathering and analysis is both crucial and difficult. Subsequent interpretation of data is then needed to pinpoint where specific phenomena (e.g., sharp bends or cracks)

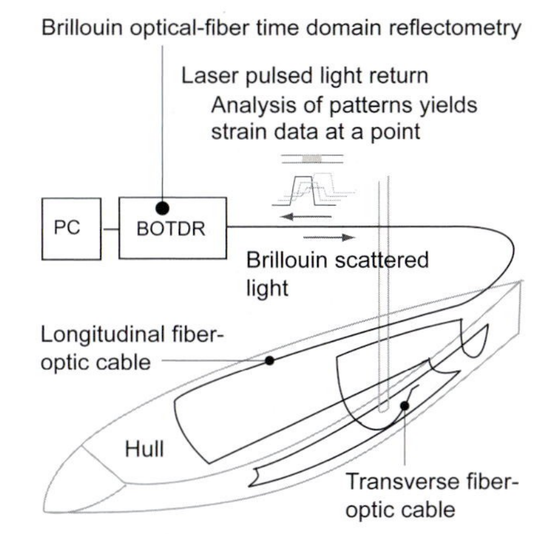

▲ **Figure 7-13** A fiber-optic system was tested for use in 'structural health' measurements for the IACC America's Cup yacht, 2000

occur. Further interpretation is needed to derive measures such as stresses from strain measurements.

Despite the apparent complexity of these fiber-optic-based damage assessment strategies, there have been many successful uses. A highly publicized case in point is the use of fiber-optic strain sensors that were installed on the yacht *Nippon Challenge*, a contender in the America's Cup competition in 2000. Fiber-optic cables were put along the length of the hull of the boat and transversely around bulkheads. The lengthwise cables were used to monitor the deformations and related flexural rigidity of the hull. The transverse cables were used to detect any possible debonding or separation that might be occurring between the hull and the bulkheads. A brillouin based system (BOTDR – brillouin optical time domain reflectometer) was used. Pulsed signals were sent along fibers. Backscattering to the original source was measured and frequency shifts analyzed. This information was analyzed to yield needed information. The installation was fundamentally experimental since it was never used in real time during a race, but it is suggestive of how these technologies may ultimately be used to improve performance. In the civil engineering and building structure area, applications include the use of fiber-optic cables embedded in dam structures, on cables of cable-stayed bridges, and on building elements. Most of these applications remain experimental or limited to monitoring highly selected critical areas (i.e., not whole structure monitoring).

A second major approach to structural health monitoring noted by Fukuda and Kosaka is based on *piezoelectric* technologies. As previously noted, strains induced by forces in piezoelectric materials generate detectable electric signals; hence their wide use as strain indicators. They can be used for measuring both static and dynamic phenomena. These devices are widely used to measure static strains at selected locations. The information obtained is thus localized. They can also be used in distributed sets and have linked data-gathering and interpretation modules.

Piezoelectric devices can also be used in vibration monitoring. They have a particularly wide dynamic range and can be used for measurements over a wide frequency range. The measurement of vibratory phenomena provides a useful way of more broadly assessing the structural health of large structures. Analyses and/or measurements of modal shapes and frequencies over time can allow investigators to determine whether damage has occurred anywhere in a structure. Sophisticated modeling techniques then need to be employed to pin-point damage locations. While lacking directness, this approach does provide a way of dealing with large structures or when critical locations are inaccessible.

There are several interesting new developments in the area of smart paint that utilize piezoelectric materials. These paints contain tiny distributed piezoelectric particles throughout a polymeric matrix. A target application area for these paints is damage detection and assessment (see Chapter 6).

Other technologies for structural health assessment include the use of *magnetostrictive* tags. Magnetostrictive materials convert mechanical energy associated with mechanically induced strains to magnetic energy, and vice-versa. Magnetostrictive tags are simply small particle or whisker shaped elements that are embedded in the base material. They are frequently the technology of choice in nonmagnetic composite materials where they are embedded in the basic matrix of the material and distributed throughout it. Measurements are subsequently taken on the magnetic flux levels (which can be measured by various probes) near the material when it is under stress. Analyses of this data can yield insights into the presence of damage. While obviously cumbersome, the fundamental notion of embedding tiny smart materials throughout a material as part of its manufacturing process is quite elegant.

Various electric resistance techniques are also in wide use. The common strain gage directly affixed to a member allows the measurement of strain via an electric resistance approach. Strains cause changes in the length and cross-sectional area of affixed looped wires which in turn cause resistance changes that can be detected and calibrated to yield strain measurements. The new approach is to utilize specific integral components of a composite material directly as the resistance element. Thus, in carbon fiber composites, the carbon fiber itself can be used as the resistance element. These approaches have been explored in carbon fiber reinforced polymers, carbon fiber reinforced concrete and carbon fiber/carbon matrix composites. Complex electrical paths develop in members made out of these materials that can be measured. Disruptions to the paths caused by damage or excessive strain affect the pathways and can be measured. The elegance of this approach lies in using the very material itself to serve a damage reporting function.

CONTROL OF STRUCTURAL VIBRATIONS

We have seen in the previous section that there are ways of assessing what damages might occur in structures. There are also ways of preventing the damages in the first place via different active control approaches that are designed to provide forces, stresses or deformations that in some way balance or offset those causing the damage, or which change the vibration characteristics of a structure to prevent unwanted and damage-causing phenomena of this kind. In this section we will look at vibration control. This is a huge topic with a long history. Various kinds of prestressing techniques, for example, have long been used to control wind-induced dynamic flutter in cable structures; or, it is well known that huge tuned mass dampers have been installed in the upper floors of tall office buildings as a way of damping the lateral motions caused by winds that both threatened the structural integrity of the building or caused occupant discomfort. This section will not cover these many well-known applications, but rather focus on newer developments associated more specifically with the use of smart materials.

The control of vibratory phenomena has been a central objective of many research and development efforts in the smart materials area. Vibratory phenomena may arise from external forces (winds, earthquakes), from machinery carried by a structure, from the human occupancy of the building, or other sources. Effects of vibrations vary. On the one hand, they may simply be troublesome nuisances that cause human irritation or discomfort. On the other hand, vibrations induced in buildings or bridges by earthquake or wind forces can potentially cause catastrophic collapses because of the dynamic forces generated in the structures by the accelerations and movements associated with vibratory motions.

These dynamic phenomena are surprisingly well understood by engineers, and many computer-based analytical models exist for predicting how specific structures respond to vibratory motions (e.g., natural frequencies, modal shapes) and the kinds of forces that are subsequently developed within them. This analytical knowledge allows designers to understand how and where to intervene in a structure in order to mitigate problems induced by vibrations. Response characteristics may be changed by altering primary member sizes and stiffness, or by redistributing masses. Other more discrete interventions take the form of the insertion of vibration *isolation* or *damping* mechanisms at different locations in a structure. These devices can change dynamic response characteristics in a controlled way, and lead to reductions in vibration-induced movements and internal forces. Thus, there are various kinds of specific isolation devices that have been developed for different applications. For vibrating machines, these devices are placed underneath the machines and serve to prevent machine vibrations from being transferred to the supporting structure. In whole buildings subjected to earthquake loadings, larger and tougher base isolation devices have been placed beneath column or wall foundations to prevent laterally acting earthquake motions from being transmitted to the supported structure. Various other kinds of damping mechanisms have been devised to be placed between beam/column connections to act as energy absorbers.

There are many conventional devices that have been developed for use as damping mechanisms or other energy absorption devices. These include passive tuned mass dampers and active mass dampers (both utilize large masses and accompanying damping devices that are typically placed at upper levels in a building), different kinds of friction dampers (e.g., bolt-type friction dampers), and hysteresis dampers that utilize low yield point steel. The design of a damping system depends on many factors, including the nature of the vibratory phenomena (especially the frequency range), the mass and shape of the structure, the use context and so forth. Of immediate importance is that of the frequency range. Many conventional viscoelastic materials work quite well for damping high-frequency modes, but are often less effective at low-frequency modes. Low-frequency modes are often extremely problematic for large-sized building and bridge structures.

Both passive and active damping devices have been devised. Passive systems directly minimize problematic vibrations by some sort of energy dissipation device, including a variety of conventional damping systems. While devices can be designed for target frequency ranges, there is little control over the action of these devices once they are installed.

Active devices are quite different in that they generally include a sensor system to detect motions, a control system and a responsive actuation system that provides some corrective action, such as applying forces or displacements in a way that minimizes unwanted vibrations or reduces amplitudes. Sensors may include traditional accelerometers and strain gages as well as some of the smart materials described below. Actuation systems may include conventional electromechanical devices or directly use appropriate smart materials. Control systems (normally based on microprocessors) that acquire, analyze and govern response mechanisms are essential and highly complex. Typically, active systems also require a mathematical model of the dynamic behavior of the structure to govern response mechanisms.

In addition to the many conventional damping systems developed for use, a variety of smart materials have also been recently used in successful ways, particularly piezoelectric, electrorheological and magnetostrictive materials.

Piezoelectrics

Piezoelectric devices have proven effective because of their capabilities for serving both as sensors and actuators. Both passive and active approaches are in use.

Passive systems assume a variety of forms, but generally use one piezoelectric device bonded to a member. One approach is often called 'shunt damping'. A piezoelectric device changes mechanical energy associated with strain deformations to electrical energy. The resulting output signals from the piezoelectric transducer are picked up by a specially designed impedance or resistive shunt, which in turn causes the electrical energy to dissipate. This results in a damping action. These devices are relatively hard to control, and are generally designed to work at specific targeted modal frequencies. 'Piezoelectric skis' provide an interesting product design example of an energy dissipating system. The problem of vibrating ('chattering') skis is well known to advanced skiers. As skis vibrate, they lift off the snow causing a loss of contact and thus a loss of control. Several companies have developed passive piezoelectric damping systems that are built directly into skis. The bending of a ski creates output electrical energy that is shunted to an energy dissipation module that in turn reduces vibration. The vibration control unit is placed just in front of the binding, where bending is maximal.

Other kinds of piezoelectric technologies include the use of piezoelectric polymers and the development of piezoelectric damping composite materials. The composite materials are intended to serve as passive self-damping surfaces. Piezoelectric rod-like elements are dispersed throughout a viscoelastic matrix. Conductive surface materials serve as electrodes to pick up output signals. These materials are proposed for use in different ways for damping. Active actuator patches are also proposed.

There are also many *active* piezoelectric vibration control devices Paired piezoelectric materials are bonded to either side of a member. One side acts as a sensor and the other side as an actuator. As the member deforms, strains are induced in the sensing piezoelectric material, which in turn generates an output voltage signal. The signal is picked up by a controller (normally micro-processor-based) that contains appropriate algorithms for analyzing input data and governing the actuator component so that it serves to mitigate vibrational movements. Signals are sent to the piezoelectric actuator, which provides the actual forces.

The active paired sensor-actuator piezoelectric devices are generally robust and can be designed to respond to multiple frequencies. Passive shunt devices are less controllable and used for relatively small structures. Their use of single elements and compactness makes them attractive. There have been efforts to make single element devices serve active functions as well, but these efforts remain largely in research stages.

Electrorheological and magnetorheological materials

A viscous fluid is rather like a semifluid. It can be thick and, according to Webster's Third, suggestive of a gluey substance. Highly viscous materials (such as heavy oil) do flow, but more slowly than do liquids such as water. A viscoelastic material exhibits properties of both viscous and elastic properties. Many conventional dampers (e.g., including many common cylinder/piston/valve devices) use viscoelastic fluids as a primary energy absorption medium. These devices are typically designed for a specific target frequency range and may not be effective outside of that range.

Many fluid materials have particularly pronounced rheological properties (i.e., properties of flowing matter) that make them ideal candidates for use in vibration control applications. Smart rheological fluids have properties that can be reversibly altered by external stimuli. Thus, the level of viscosity of electrorheological (ER) materials can be varied by electrical stimuli, and that of magnetorheological (MR) materials can be altered by varying the surrounding magnetic field. Since the viscosity of these materials can be altered, damping devices utilizing them can be designed to be tuned to varying frequency ranges. Since the smart materials used are based on electrical phenomena, and respond very quickly, viscosities can be controlled quite well and be programmed to respond to varying conditions obtained from sensory data and/or analytical vibration models.

The physical make-up of these kinds of smart dampers varies widely. In some systems either electrorheological or magnetorheological fluids may be encased in laminates that are applied to structures in different ways, and which are connected to a control microprocessor. Varying the electrical or magnetic stimuli causes the laminate to stiffen or become more flexible, thus altering the vibratory characteristics of the base structure. While still far off, the idea of making a whole laminated surface with inherent damping capabilities is not without feasibility.

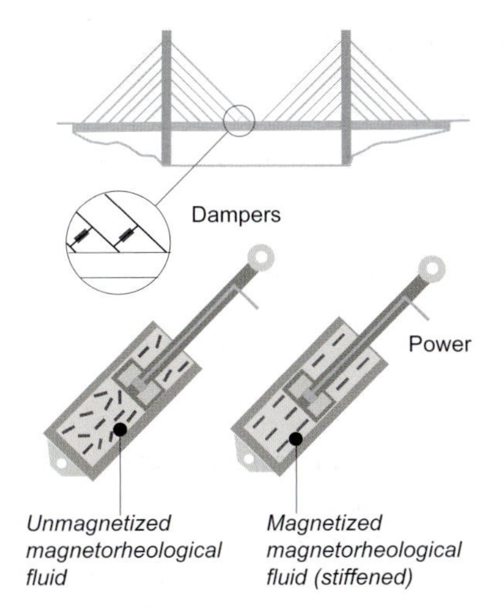

▲ **Figure 7-14** Applying a controlled current creates a magnetic field that causes the viscosity of the magnetorheological fluid to vary and damp out unwanted cable vibrations

Various kinds of electrorheological devices have been proposed for a wide range of automotive and consumer products as well. Conventional shock absorbers in cars, for example, may soon be replaced with smart shock absorbers based on smart fluids that can in turn be controlled in real time to provide improved rides.

Base isolation technologies have proven to be one of the great success stories of in the development of damage-reduction techniques for buildings and other structures subjected to earthquakes. These devices are placed at the bases of structures and isolate the structure from ground accelerations, thereby minimizing forces developed in the supported structure. Their effectiveness has been repeatedly demonstrated. Conventional base isolation systems assume various forms, including lead-rubber systems. Another approach uses elastomeric bearings in conjunction with a damping mechanism. In the latter approach, magnetorheological fluids have been introduced to make the system smart and controllable. Smart fluid dampers can potentially be controlled in real time based on data obtained from sensors that measure ground and building motion. Related analytical simulation models can then be used effectively to modulate the behavior of the smart fluid dampers to best optimize their performance.

Other materials

Virtually any material that undergoes reversible shape or stiffness changes could conceivably be used for vibration control, since any change in these parameters would influence overall vibration characteristics. The use of shape memory alloys, for example, has been explored in connection with vibration control; particularly for small-sized applications. Their slow response characteristics, however, make them suitable for only limited applications.

CONTROL OF OTHER STRUCTURAL PHENOMENA

Most of the discussion thus far has focused on vibration control, since this is one of the current major application domains for smart materials. Many other structural phenomena, however, can also be controlled. Engineers have long sought to control problematic static deflections of beams or larger frameworks via various kinds of static prestressing techniques. Thus, a reinforced concrete beam might have embedded prestressing cables that cause the beam to camber upwards to offset downward deflections induced by external loadings. There have been attempts to vary prestressing forces in response to the level of the externally induced deflection. Truss members have had actuators built into specific members to alter force distributions and related deflections.

Various shape-changing smart materials could possibly be used in many of these applications. The active paired sensor-actuator piezoelectric systems described previously for vibration control, for example, could be used to control beam or framework deflections by developing forces that balance or counteract those generated by externally acting loadings. These systems have also been experimentally used to control incipient buckling of slender members. Shape memory materials could also be used. Experiments have been made with devising active truss structures for large flexible space structures via the use of piezoelectric technologies.[11]

In most of these applications, however, the members controlled are currently small in size as are the magnitudes of external loadings – at least in comparison with the very large members and loadings typically encountered in building and bridge construction. For active control, the latter often require extremely large forces to alter their behavior that exceed what is currently easily feasible with typical smart materials. Nonetheless, there have been interesting experiments that suggest a bright future for the active control of structures.

Notes and references

1 For a more complete discussion of the origins and development of HVAC systems over the last century, see D.M. Addington, 'HVAC', in S. Sennott (ed.), *Encyclopedia of 20th Century Architecture* (New York: Fitzroy–Dearborn, 2004).

2 Cited from Banham, Reyner (1984) *The Architecture of the Well-Tempered Environment,* 2nd edn. Chicago: The University of Chicago Press, pp. 292–293.

3 Cited from Bell, J.M., Skryabin, I.L. and Matthews, J.P. (2002) 'Windows', in M. Schwartz (ed.), *The Encyclopedia of Smart Materials,* vol. II. New York: John Wiley & Sons, pp. 1138–1139.

4 Seeboth, A., Schneider, J. and Patzak, A. (2000) 'Materials for intelligent sun protecting glazing', *Solar Energy Materials & Solar Cells,* 60, p. 263.

5 See Bell *et al.*, 'Windows'.

6 For a more complete discussion see D.M. Addington, 'Energy, body, building', *Harvard Design Magazine,* 18 (2003).

7 Cited from 'Fiber-optics: theory and applications', *Technical Memorandum* 100, Burle Industries, Inc.

8 Laboratory results have been much higher, reaching 30% for Gallium Arsenide cells, and are predicted to reach 40% for Quantum Well cells. See the chapter 'Present and proposed PV cells' in R. Messenger and J. Ventre, *Photovoltaic Systems Engineering* (Boca Raton, FL: CRC Press, 2000).

9 Messenger and Ventre, *Photovoltaic Systems Engineering,* p. 182.

10 Much of this section on structural health monitoring is adapted from T. Fukuda and T. Kosaka, 'Cure and health monitoring', in *Encyclopedia of Smart Materials*, vol. I, ed. M. Schwartz (New York: John Wiley & Sons, 2002), pp. 291–318.

11 Bravo, Rafael, Vaz, F. and Dokainish, M. (2002) 'Truss structures with piezoelectric actuators and sensors', in M. Schwartz (ed.), *The Encyclopedia of Smart Materials*, vol. II. New York: John Wiley & Sons, p. 1066.

8 Intelligent environments

We have materials that act, and machines that can buzz around us. Someone in Alaska can follow every move of a bird migrating from Canada to Costa Rica, while *not* watching her kettle, which, due to a shape memory alloy, will automatically shut off. As intelligence and self-actuation have been embedded in the objects and processes around us, providing us with unprecedented control of our environment and knowledge of our surroundings, they have also given others open entrée and access to our private lives and activities. Even though material scientists developed these materials, engineers built the technologies, and programmers wrote the code, architects and designers are left with both the responsibility and the challenge of integrating these developments into daily life. We design the environments that people live in, and the places that they go to. We are now left with the dilemma of just what it is that we are designing.

Do we design buildings, cities and landscapes, or do we design spaces, places and views? Do we design to meet an individual's needs or to physically express a social condition? Do we design for beauty or pragmatism? Many would argue that we do them all. Whatever the reason, and it may well be all of them, the result is almost always a physical *thing*. It will be comprised of components that can be specified, weighed, measured and assembled. With human occupation, this static collection of physical things becomes an environment – one that might delight, or one that might repress. We know that Bernini and Le Corbusier are capable of creating environments that transcend the bounds of their physical surfaces, while our local financier and developer most likely are not. As such, we have traditionally considered that great environments were those defined by and limited to the work of the best designers. The rest of us could only hope that one day we too could join their ranks.

Nowhere, however, is the problem of context more profound than in determining the definition of the environment surrounding us. While architects and designers consider the environment to be an unquantifiable result produced by a quantifiable physical object, scientists and engineers consider the environment as a quantifiable process that determines the resulting nature of a physical object. Do we bring smart materials into the design field as we have brought other materials, and treat them as part of our design palette of physical artifacts, or should we cede our design process to the visions of the technologists and scientists?

In this chapter, we will examine the differing views as to the constitution of an environment and furthermore as to what makes it intelligent. We have briefly opened this discussion in Chapter 1 (see Figure 1–8). Here we continue the exploration. Each side of the debate has been couched in the traditional paradigm of the differing professions: materials and technologies are the determinants of the designed objects, and thus are design objects in themselves, or materials and technologies derive all the systems that determine the environment, and are thus subordinate to their function. We will then try to find an appropriate middle ground which respects the integrity of the design process while opening it up to radically different approaches.

8.1 The home of the future

Design has never been content with the status quo, and while we might look to engineers to produce the things we will use in the future, it is the designers who have always told us what that future would look like. If we briefly recount the history of the future, we will begin to see that it was frequently defined by the materials and technologies from which it would be constructed. While utopian projections and ideals have populated the practice of architecture for centuries, it was not until the advent of modernism that we began to see a concerted effort to speculate on the future. Indeed, in 1914 a group of young Italian architects published a 'Manifesto of Futurist Architecture' which proclaimed that 'Futurist architecture is the architecture of calculation, of audacity and simplicity, the architecture of reinforced concrete, of iron, of glass, of pasteboard, of textile fibre, and of all those substitutes for wood, stone, and brick which make possible maximum elasticity and lightness.'[1] Manifestoes abounded during the first half of the 20th century, and while all of them were inherently concerned with social conditions, and thus an architecture that rejected the frivolity of the bourgeoisie, they relied on form and material to communicate their message physically. Concrete was perhaps the most favored material of the modernists as it represented a clean break from the materials of the past, and its proletarian and massive presence was in concert with the socialist ideals that architects of the early 20th century generally espoused.

In regard to technology, Le Corbusier defined the house of the future in 1920 as 'A Machine for Living', and 9 years later,

Buckminster Fuller published the drawings of his 'Living Machine', the Dymaxion house. Both designers took their inspiration from the nascent industries of automobile and aircraft assembly, tying into the aesthetic as well as the principles of mass production. Accompanying Fuller's drawings of his Dymaxion house was but one of the many didactic pronouncements that filled architectural publications of the day:

> having once freed our minds of the customs and traditions that have bound us since the days of the earliest shelters, we can attack this dwelling problem just as we would attack the problem of building some other device or piece of machinery that had never before been made.[2]

As the 20th century progressed the ideological symbolism of new materials and technologies was steadily replaced with an iconographic utility. Keck and Keck took the inspiration of aircraft manufacturing a step further than their predecessors by including an aircraft hangar on the ground floor of their 1934 'House of Tomorrow'. The Aluminaire, built exclusively for the Museum of Modern Art's first architectural exhibition in 1932, was intended as a demonstration of the potential of aluminum both as a façade material and for the structural system. Houses that demonstrated a company or industry's latest material began to proliferate: the Masonite House, the Stran-Steel House, the Armco-Ferro Enamel House. As World War II was winding to a close, the US government supported an enormous enterprise that they hoped would solve two

▲ **Figure 8-1** Monsanto's 'House of the Future' opened at Disneyland in 1957. The house, and all its furnishings, were fabricated of plastic

problems: the need for housing for the returning GIs, and a continuing market for the steel industry after the war. $38,000,000 later, the first pre-manufactured Lustron home emerged. True to its initial premise, it was all steel, *all of it*, exterior and interior. (Recently, the elderly occupant of one of the last remaining Lustron homes remarked that it was easy to clean the kitchen, she only had to hose it down.)

By the 1950s and 1960s, the home of the future was either a blatant marketing tool or a folly. Monsanto's Home of the Future showcased many of the company's plastic products as did General Electric's Concept House. Coinciding with the shift from the industrial age to the information age, homes like Xanadu, with its amorphous polyurethane shell, marked the end of an era in which the material and the house were one and the same. Indeed, when Philips Electronics unveiled their predictions for the Home of the Near Future in 1999, they described it as looking 'more like the home of the past than the home of today.'[3]

The intimacy between materials and architecture has always existed, but the 20th century represented a time when materials and technologies were given additional roles – ideological, didactic, iconographic and the very pragmatic one of saving an industry. Materials continue to be chosen not so much for how they perform, but what they connote. As such, smart materials and new technologies pose a dilemma, because at the scale of their behavior, they have very few connotative qualities.

8.2 From the architect's view to the technologist's view

So while architects try to wring a unified visual aesthetic from these materials, engineers and computer scientists are looking to build a unified system of the technologies. Their version of the smart home or house of the future is a formless one, the physical aspects of the architecture recede into the background as performance is foregrounded. In its earliest manifestation, the 'All Electric House' of 1953 allowed the occupant to flip a switch in the bedroom to turn on the coffee maker in the kitchen. Following this model, 'smartness' was equated with automation, with lights turned on at twilight, the lawn watered when dry, coffee made in the morning, and so forth – even the dog could be let outside or have its food bowl filled at certain times. If this is smart, then what of the classic thermostat? This ubiquitous and simple technology is responsible for the thermal management of almost every building, but rarely does anyone think about it as being an 'intelligent' device.

Are home automation approaches 'intelligent environments?' At present, most automation approaches of the type just described are not uniformly considered to be 'intelligent environments', nor are they considered to be 'smart'. Clearly, the prevalent assumption is that simple 'control' or directed operation of something by a specific technology (a timer turning on the coffee maker in the morning) is somehow not considered 'intelligent'. Are we then to interpret home environments filled with robotic devices that vacuum the floor or clean the bath tub, in the style of *The Jetsons*, to be proposals for 'intelligent environments'? A group of computer scientists clearly believed so when they presented their 'intelligent inhabited environment', in which an 'Intelligent-Building [is] inhabited by a variety of agents ranging from mobile robots, embedded agents to people some of which carry smart wearable gadgets, [and all] cooperate together'.[4] The interest in tying the 'gadgets' together into a seamless and single system has clearly expanded into including humans as yet another 'device' that can be identified, tracked and maybe even controlled. Already, the human body is 'tethered' to the digital world through the many wearable and portable devices each of us routinely carry. And, already passé, are the infrared business cards that we could beam to strangers through our PDAs, as the technology has progressed beyond. We are quickly approaching an era where everything will be cataloged and outfitted with implanted sensors or RFID tags. Much of this is argued on merits of human welfare and safety – an implantable tag will tell paramedics what we may be allergic to if we are unconscious, just as how these tags identify a lost dog's owner. Nevertheless, once a human joins in with the technology, he or she becomes but one of the digital devices that are factored into a larger system. For example, RFID systems now monitor customer behavior, determining where they pause and what they ignore, and can tie that into a 'smart' inventory system that proactively adjusts the merchandise mix.

More recently there has been a lot of fascination with so-called 'information-rich' environments that are clearly positioned within today's preoccupation with communicating, receiving, displaying, interpreting and using digitally based information. We have systems that identify specific human presences or interpret gestures, and then try to use this knowledge as a way of easing or improving their work environments. We also have 'tangible' media that, by transforming digital information into physical references, attempt to bridge the digital and real worlds. Many systems are currently being proposed in which, through a 'tag', individual user's preferences are communicated to the HVAC system so that the conditions in the office can be tailored for each occupant. And perhaps the strangest conflation of the two worlds occurs in ubiquitous computing, in which information is communicated through everyday objects: for example, a picture frame that, when picked up, reminds you of the birthdays and preferences of your relatives in the photo.

Just as we recognized that the design world envisioned the future as material-centric, the engineering world and computer science worlds envision it either as inhabited by smart products or as subordinate to an overarching communications structure. These are not insignificant differences. And while there is no universally accepted meaning to the term 'intelligence', there are instead as many ways to characterize it in buildings and products as there are in humans. In the following sections, we will first construct a classification system – albeit a simplistic one – to help in understanding the multitude of meanings attached to the phrase 'intelligent environments'. We will argue that within these meanings, the specific concepts of 'smart systems', as discussed in the previous chapter, do indeed have a place. The aspirations of intelligent environments, however, are higher, as they must operate in multiple contexts and simultaneously interact with the transient behaviors and desires of humans.

8.3 Characterizations of intelligent environments

For designers and architects, many of these advancements in 'intelligence' are peripheral, even insofar as we appreciate their rapid and continuing absorption into our daily activities. Few of us are capable of running a design practice without full digital knowledge of our design product, and soon we will be unlikely to take a road trip without a GPS and personalized navigation system. The nature of public and private, once the domain of architects, has flip-flopped as personal communication devices keep us connected at any time, any where. So where do we fit in to this debate on the 'intelligent environment'? Can we move beyond fetishization of the gadgets, and get over our preoccupation with showing off the advanced materials in a purely provocative manner?

In this section, we are going to begin to develop our own characterization of intelligent environments. Rather than presuming an overarching hierarchical structure predicated on the individual as either being a servant of or being served by, we are considering that 'intelligence' is fluid and contingent. Many meanings may emerge, combine and recede depending on ever-changing human and environmental circumstances. As such, when we re-examine our basic hierarchy of material and environments originally presented in Figure 2–4, we begin to become aware that these are not prescriptively sequential, but are indeed simultaneous. Our goal is not to prescribe the best one or the ideal, but to enable a fluid human–environment interaction.

If we look at the broad meanings explicitly or implicitly present in the term 'intelligent environment', three large groupings can be differentiated:

- *Environment characterizations.* One stream of meaning revolves around the *human occupation and use of spatial environments*; including the enhancement of the processes that respond to and support human living, lifestyle and work needs. A related stream is centered on characteriza-

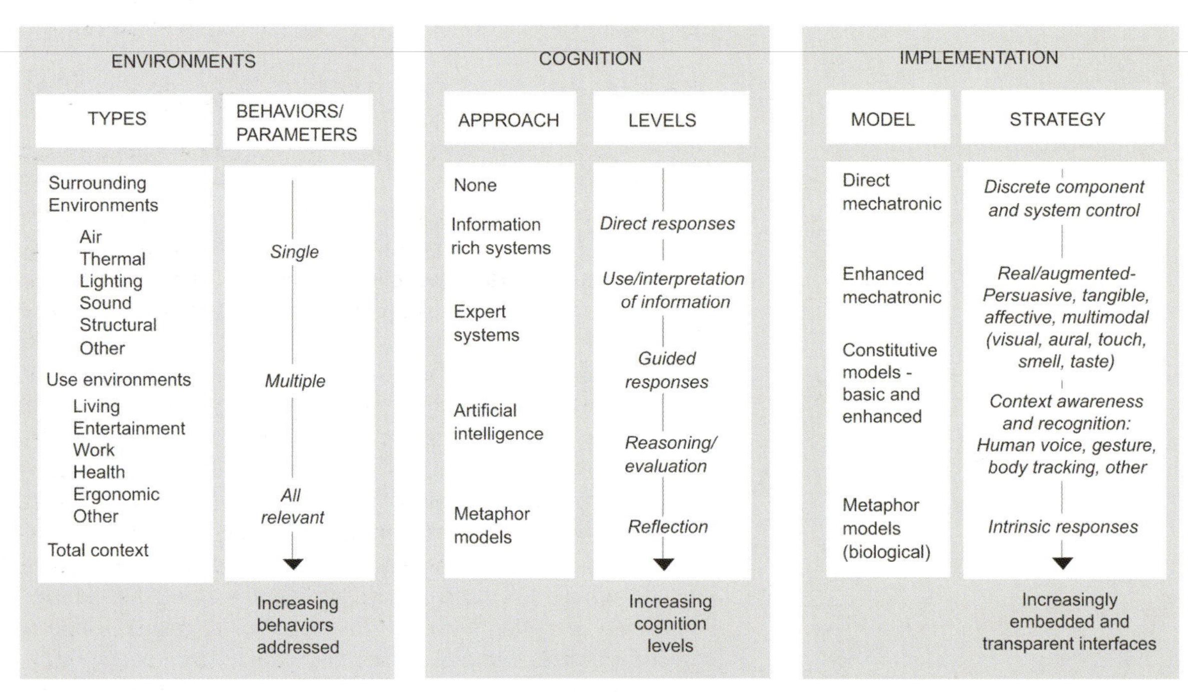

Figure 8-2 Intelligent environments can be broadly characterized in terms of their functional descriptions, the ways that smart behaviors are embedded in the environment or the ways they are controlled, and the general cognition level associated with the different behaviors

tions of the *physical environments* that in one way or another support and enhance the human occupation and the use of the space.

- *Cognition characterizations*. A third stream deals with the extent to which cognition-based processes are embedded in the environment or use setting. Included here are approaches that engage human emotions, thoughts and cognitions occurring within the environment.
- *Implementation characterizations.* Another stream of meaning is centered on the ways in which these enhancements are *invoked, operated and controlled.*

While these streams might appear to be human-centric, they are as applicable to society and community as they are to the individual. As a result, even though each person may have 'control' of their surroundings, that control is always placed within greater contexts beyond the domain of the individual environment. Also implied is that the means by which actions ultimately occur is via the use of computer-assisted embedded technologies and/or smart materials. We will see below, however, that none of these four broad characterizations alone is adequate to convey a notion of intelligence in an environment, nor do they mutually exclude other ideas.

USE AND ENVIRONMENT CHARACTERIZATIONS

One major approach to intelligent environments centers on the human occupation and use of the spatial environment, and thus aids and enhances processes that (a) respond to and support human living, lifestyle and work needs, and (b) enable the general interactions between humans and their physical environments (including ergonomic responses and others).

Currently, many approaches being explored in this arena are 'information-rich' environments that seek to enhance and support human use and activities through the provision of specialized information. In simple approaches, information floods the spatial environment, while more sophisticated computationally based technologies involve specific targeted information to an identified person or group of individuals. The 'information' can be quite literal or can be broadly construed. At the primary level, the involved technological actions may be invoked by direct human initiation. A far more sophisticated approach, however, is via the use of different types of sensor systems that detect and track human presences and actions, including the identification of specific individuals, and that can interpret specific and recognizable signs of their intentions. A whole host of interesting technologies have been developed to aid in meeting these objectives, ranging from relatively simple tagging systems all the way through to facial and gesture recognition methods.

In terms of objectives, the support of processes that respond to and support human living, lifestyle and work needs can range from the pragmatic to the sophisticated. There are many aids that can be imagined to support basic living functions, even including the myriad of devices that desperately seek to organize the user: schedules, reminders, etc. More interesting are those that have aspirations to support and improve living processes for particular groups in need, such as the elderly or the disabled, and/or with respect to specific objectives, such as health care. Thus, we find initiatives to allow 'aging in place' to occur within the home, or provide the 'intelligent bathroom' for aiding in personal health monitoring and inspection, or other various assistants for diabetics and others.

In terms of processes that purport to support lifestyles and work-processes, we are literally inundated with varying initiatives, most of which fall into the category of information provision. For lifestyle support, applications such as entertainment are seemingly introduced almost daily, e.g., enormous screens that provide 24-hour wrap-around movies or other entertainment tuned to user preferences. This world can extend into exercise regimens, culinary experiences, and so forth.

Applications that support work processes are usually more conceptually interesting. Many are again typically based on information provision. Here we have proposals for rooms with various technological paraphernalia that enable a stockbroker to better accomplish his or her activities by providing real-time data on current situations. Other work process aids have different orientations. We have tracking systems for locating and communicating the status of different 'objects' or tools that individuals might use in their activities, e.g., various kinds of location, status and inventory systems developed for keeping track of critical devices within a hospital setting.

More sophisticated approaches increase the extent to which enhancements of living and work processes are simply information-communication oriented, and become increasingly based on the interpretation and utilization of the information as a means of improving living and work processes. In most current applications, these enhancements occur via computer assistance.

Related to use characterization, the surrounding environment characterizations in some way directly support and enhance the human use of space by maintaining the quality and determining the behavior of the surrounding physical environment. Some are obvious to us, such as management of the thermal, luminous and acoustic environments, and some operate below our normal plane of observation, such as dynamic structural control. Here there are a wide range of approaches that relate to the following: (a) the air, thermal, sound and lighting environment; and (b) the ability of the physical environment to continuously provide a safe environment under all conditions, including adverse ones (approaches that enhance the performance of structural systems, for example, would fall here).

Many, but not all, of these approaches currently deal with various detection, monitoring and control actions that are based on different kinds of sensor–actuator systems. There are, for example, broad ranges of technologies for monitoring and controlling the surrounding air, thermal or lighting environment within buildings that directly utilize sensor–actuator systems of one type or another. The same is true for structural systems that have sensor systems that detect earthquake-induced ground motions and then cause some type of response to occur, such as generating a damping action. Many sensor–actuator systems are highly sophisticated, others are relatively simple – e.g., a motion sensor and related mechanical actuator that causes a door to open as a human approaches. Should we consider these latter simple systems worthy of the term 'intelligent?' Today, we consider them unremarkable – but a few decades ago this was the dream of the future.

For our purposes here, a review of the literature suggests that the term 'intelligent' is widely used in the broad connection of monitoring and control, but our use of the term extrapolates beyond simple sensor–actuator systems with respect to (a) the complexity and or meaningfulness of the actions or phenomena to be controlled (with the clear implication that it is something that has not successfully been done before), (b) the level of sophistication of the responding technology, (c) the use of computationally assisted operations and controls and (d) the extent to which the operations and controls involved are cognition-based and transparent (see next sections). Thus, in this positioning of the use of the term 'intelligent', the 'smart environments' previously discussed may or may not be considered 'intelligent'. For example, a sophisticated 'structural health monitoring system' that assesses the overall performance of high-end sailboats that is based on embedded fiber-optic technologies might be described as 'intelligent' if the information obtained is not an end unto itself, but is then used in some way to control the overall actions and performance of the sailboat in a cognition-based way. Other 'smart environments' (thermal, air, etc.) could be thought about similarly.

In this discussion, we will consider environments that involve the *detection, monitoring and control* of a *single* behavior or action via embedded computationally assisted technologies to be a valid but lower level characterization of an intelligent environment. The more the system exhibits the cognitive behaviors described in the previous section, the more the environment may be considered 'intelligent'.

A related but more sophisticated environment would be when *multiple behaviors* and their interactions are considered. We have encountered before the significant differences between dealing with single behaviors versus multiple behaviors and their related interactions (see Chapter 5).

In single and multiple behavior instances, the presumed operations and control model is either the 'mechatronic (mechanical-electronic)' or 'constitutive' model (see implementation characterizations described below), although more advanced means are possible. Clearly, single and multiple behavior characterizations could be further refined by considering the operations and control model used.

As we think more speculatively about these kinds of environments, the interesting question arises about how current approaches might evolve over time. For example, might not the now traditional role of the physical boundary in a building (e.g., a wall) that serves multiple functions – as a thermal barrier, a weather barrier, a light modulator, etc. – be reconsidered and non-coincident phenomenological boundaries created instead? Here a primary concept of interest emerges around the issue of *selectivity of response or action.* A closely related issue is that of the value of smart materials and other technologies to *dis-integrate* certain behaviors or actions that currently occur within a building or other environment at system levels or truly macro-scales, and to replace them with multiple *discrete* actions. We have encountered this concept before in our earlier discussions of smart actions and smart assemblies (see Chapter 7). Let us think about a common human need in spaces – that of an appropriate thermal environment – and revisit a speculative example cited earlier. Right now, most systems seek to provide for human comfort by heating or cooling entire room-level spaces within buildings. Might not there be ultimately found a way to selectively condition only the local environment *immediately surrounding an occupant*, instead of whole rooms? The potential benefits of these approaches are manifestly obvious and could be discussed at length. Here, however, the important message is that this is an example of *selectivity*. It also suggests a *discrete and direct* approach to maintaining environments. Many other similar strategic interventions could be noted. In this discussion, we will define this level of aspiration to be higher-level intelligent environment characterization.

Cognition-based characterizations

The term 'cognition' is used here in its common-sense meaning of an intellectual process by which knowledge is gained, utilized and responded to. Here we also liberally include all processes that engage the human emotions that occur within the environment, as well as thoughts and cognitions. Clearly, this world is elusive and hard to define, yet these processes are ultimately a defining characteristic of the concept of 'intelligence'.

We begin by considering varying levels of cognition-based processes. On the basic level, it is evident that 'information rich' environments of the type just discussed in the last section and those that are in some way specifically designed to be 'cognition-based' are not the same, but defining exact distinctions is difficult. An information-rich environment is one in which relevant data or other information is provided to a user in a highly accessible way. While information may be provided, it does not necessarily follow that it can be effectively utilized by a user. Still, an information-rich environment is generally a necessary precursor to a cognition-based process.

The problem with the human use of information has been addressed many times. One of the most significant issues is simply the staggering quantities of information now available for even the simplest processes. There are currently many workable computer-enhanced systems that have been developed to aid an individual in coping, understanding, and effectively utilizing complex information sets; and, in so doing, directly support or aid a myriad of creative activities, work and so forth. Some of the first explorations in this area were called 'knowledge-based' or 'expert' systems. Expert systems essentially codify best practices into a set of rules that can be used for sifting through all of the data and then advising a human operator on the historically best responses to a specific situation. The knowledge is contained in the rules, and the intelligence belongs to the operator. A common example of where expert systems have been widely utilized is in the medical field for diagnostic applications.

Fuzzy logic adds a dimensionality to expert systems. Whereas expert systems match current conditions to past conditions that have a known 'best' response, fuzzy logic additionally maps current data to multiple sets of data to produce more than one possibility. This approach is an attempt to shift some of the intelligence from the operator to the system so as to bring in some of the instinctive reasoning that allows new and possibly even better responses than in the past. Both of these systems are considered 'supervised' in that a human still makes the final decision. We must be aware, however, that these systems do not control activities, they simply provide the guidance for the more conventional control schemes (i.e. feedback, feed-forward) that still depend on the mechanical behavior of actuators to enact the response.

These approaches aid a user in understanding and utilizing a complex information environment. Some extend into more advanced modes that contain algorithms that mimic human decision-making. The addition of capabilities of this type is a significant step towards making systems that are truly 'cognition based'.

A related but even more sophisticated approach that is gaining currency is when the involved technological actions actually *anticipate* human needs or interests and are already working by the time the human action actually begins. This notion of 'anticipation' is an interesting one. It ties back in to the earlier discussion of 'intelligent' behaviors in Chapter 1, where the notions of abilities to understand or comprehend were suggested as a characteristic of an intelligent behavior. In order to anticipate needs, it is clearly necessary to understand or comprehend a complex situation. The idea is interesting and reflective of developments in the realm of what has traditionally been called 'artificial intelligence'. This is a hugely complex field with its own nuances of what is meant by the term. Here we accept it in its most general form in relation to its being a defining characteristic of a cognitively based advanced use environment.

Artificial intelligence is a generic term that has been used to refer to any information-based system that has a decision-making component, regardless of whether that component is advisory, as in expert systems, or is part of an unsupervised neural network that is capable of extrapolating into the unknown. Today, however, the term is more frequently used in relation to Artificial Neural Networks (ANN). Modeled after the human brain's neural processing, ANNs are designed to be capable of 'learning'. These networks contain vast amounts of data that are sorted and put through an exhaustive trial and error pattern recognition testing that is known as 'training'. Once trained, an ANN has the 'experience' to make a 'judgment' call when out-of-bounds data are encountered or unprecedented situations arise. Each level in the development of AI has progressively reduced the human participation in the real-time activity of decision-making.

As we move down the path of increasing expectations of what we ultimately want to find in a spatial environment that is deemed 'intelligent' with respect to cognition processes, we find that not only is the capability to understand or comprehend something important, but the potential power to *reason* becomes an enticing goal. Here we enter into the world of passing from understandings of one state (or propositions about it) to another state which is believed to follow directly from that of the first state, i.e., an ability to make *inferences* that in turn govern responses. Again, the term 'artificial intelligence' is currently best suited to describing these kinds of activities, but even yet more demands are placed on this still emerging and evolving field to provide reasoning capabilities as a yet more advanced form of intelligent environment.

Are there more expectations about what we might want to ultimately find in an intelligent environment in this connection? Perhaps at some point an environment might ultimately have a capability for enhancing the powerful human capability of evaluation, and then perhaps even reflection. The power of reflection is one of the most fundamental of all signifiers of human intelligence. Can our environments enhance this power? We remain largely in the domain of speculations about the future here. In the accompanying figure, we have noted a classification placeholder for environments that might be ultimately developed to enhance *evaluation and reflection* powers and other high human aspirations.

Implementation characterizations

The preceding characterizations largely focused on objectives and goals. The question of how suggested enhancements are *invoked, operated and/or controlled* – or we might use the term 'interface' in this connection – remains a large issue that was only briefly touched on in the discussions above (through references to 'sensor–actuator' systems and the like).

Ways of *invoking* the operation of an action include the wide range of sensors and other technologies already previously described. They may range in complexity from simple sensors through various forms of sophisticated human tracking, and gesture or facial recognition systems. Within the general understanding of an 'intelligent' room, these devices are generally embedded in the environment in a way that is largely invisible to the user. It is assumed that most actions are automatically invoked, albeit in some situations the need and desirability of human initiation or overrides is clearly important (as a trivial example, who has not, at one time or another, wanted to cut off one or more of the automated formatting aids found in word processing programs that purport to help one write a letter?). Ideally, the user would also not need to be in any particular location in the room or environment to generate an action.

The ways of *operating or controlling* the actions that occur within an intelligent room are difficult to easily summarize. The discussion in Chapter 5 provides one immediate way of characterizing elements or components that make up intelligent environments from this perspective. Recall that five major ways of invoking, operating and controlling complex systems were discussed, including:

- *The direct mechatronic (mechanical-electrical) model:* In this basic approach, a sensor picks up a change in a stimulus field, the signal is transduced (typically) and the final signal directly controls a response. This simple model describes many common sensor–actuator systems, including common motion-detectors that switch on lights, and so forth.
- *The enhanced mechatronic model:* This model builds on the simple mechatronic model by incorporating a computational environment that allows various types of operation and logic to be incorporated in the system. This computational model may be conceptually simple, as is the case with a host of devices that are linked to microprocessors that execute many kinds of programmed logic functions, including the sequencing of responses and various kinds of 'if–then' branches. Alternatively, they may be much more complex to the extent that the computational model may constitute a knowledge-based system of some type or lay claims to artificial intelligence.
- *The constitutive models*: These models are used in connection with property-changing smart materials – see Chapter 4), in which an external stimulus causes a change in the properties of a material, which in turn affects the response; and with energy-exchanging smart materials (see Chapter 4), wherein an external stimulus causes an energy exchange in the material, which in turn affects the response. *Enhanced constitutive models* are an extension of the models just described wherein a computational environment is built into the system to allow for various types of operation and logic control. As previously noted, the

computational model may assume varying levels of sophistication from the simple to highly complex knowledge-based approaches. Interfaces become more transparent and embedded.

- *The metaphor models:* This curious title is used here to describe a wide variety of models that are in one way or another based on some metaphor of how a living organism works. Here the stimuli, sensory, response and intelligence functions are totally interlinked and embedded. Even here there are levels, since many stimuli-response functions are largely instinctual and seemingly demand little from the intelligence end, while others engender a thoughtful response. In addition, neurological models and other highly complex systems are considered.

Within these general models are many technologies of varying sophistication. At the advanced level, there are virtual and augmented reality systems. With augmented reality systems users can see and interact with real world environments that have been enhanced by various information displays and simulations of phenomena or events. These systems can provide multimodal environments that engage basic visual, aural, touch, balance, smell and taste sensations. We also have persuasive, tangible, affective and other approaches. There are recognition and other technologies for context-awareness; including basic human body tracking, facial, voice and gesture recognition. These and other fascinating emerging technologies – beyond the scope of this book to explore in detail – show promise in making the human–environment interface both robust and, potentially, largely transparent to the user.

In current practice, most of the characterizations noted above are most clearly applicable to either single behaviors or to multiple behaviors that are used in relation only to the elements or components that make up larger systems. Situations become much more complex when whole environments are considered. In the simplest scenario, a total environment can be envisioned as consisting of many single behavior elements or components that are considered to act in an *essentially independent* way – the action or response of one does not affect others. This is a common approach in current implementations of intelligent room environments. Multiple behavior elements can also act independently of one another.

A more sophisticated and engaging scenario, however, is when there are single or multiple behavior elements that both interact with one another and mutually influence their

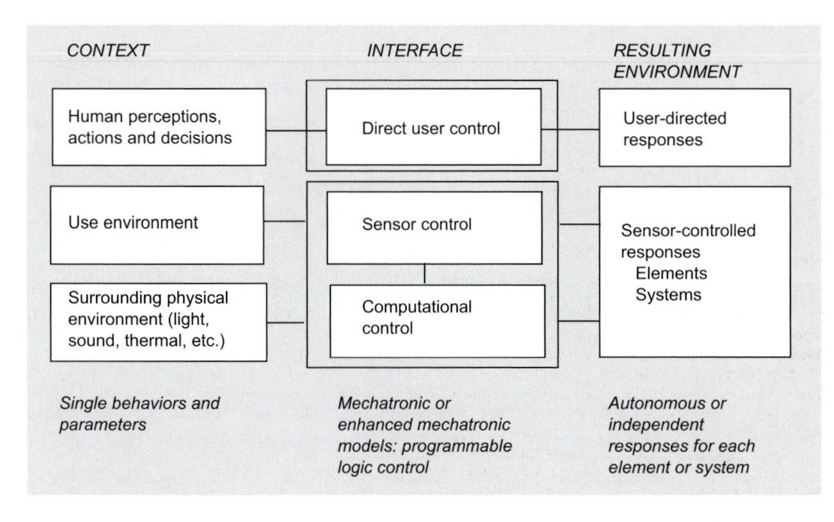

Typical current 'smart room' approach

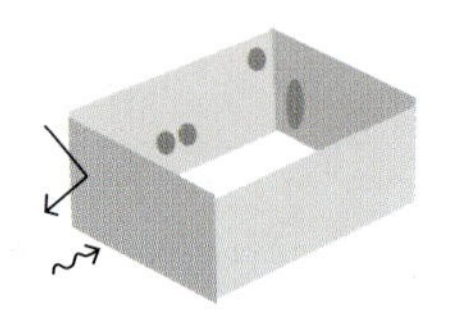

Non-embedded interfaces and sensor/actuator elements

Current 'smart room' approaches using enhanced mechatronic models (see Chapter 5 for a discussion of control approaches)

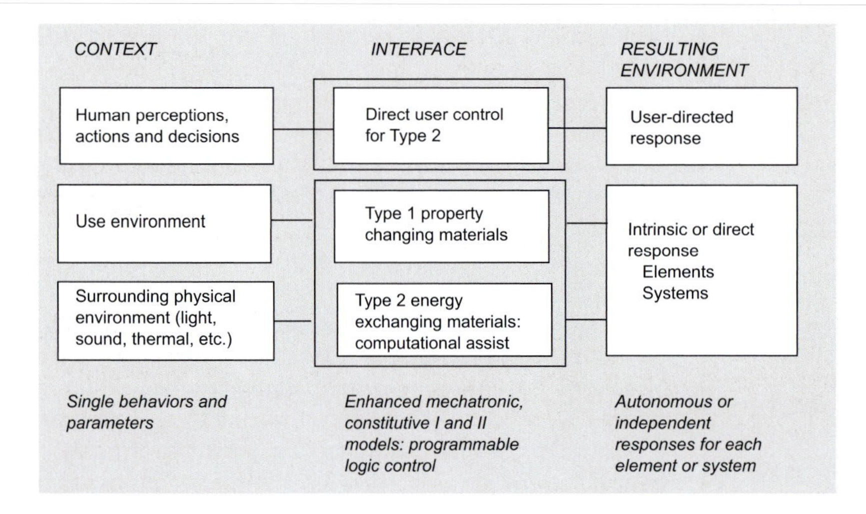

Current smart environment approaches using Type 1 and 2 smart materials

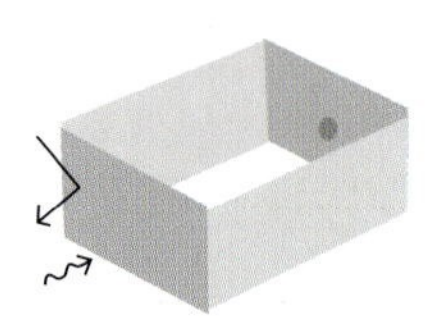

Autonomous embedded sensing and response elements acting intrinsically or directly

Current approaches to using smart materials in making smart environments via enhanced mechatronic, constitutive I and II models.

▲ **Figure 8-3** These four diagrams illustrate past, current and future approaches to the design of intelligent environments

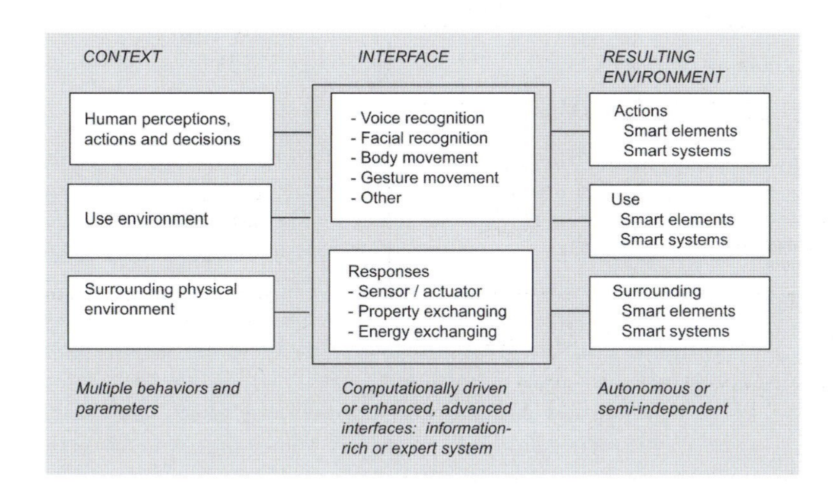

Current intelligent environment approaches

Embedded interfaces

User-centered environment

Current 'intelligent environment' approaches.

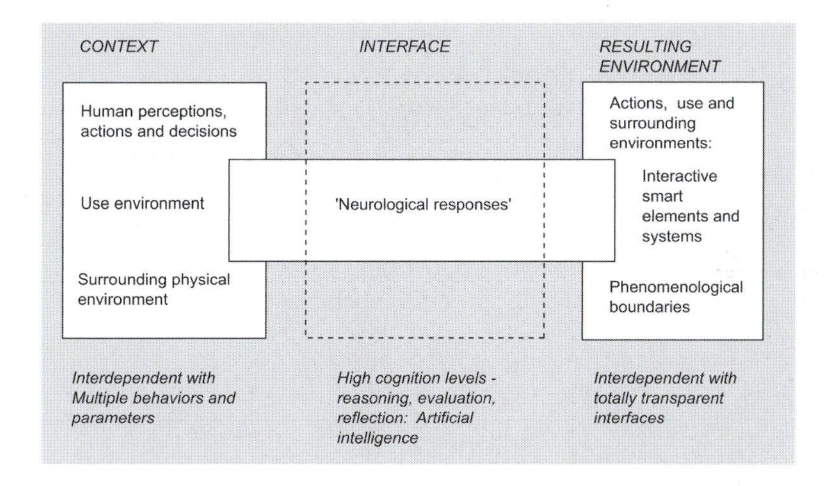

Transparent interfaces

Thermal space boundary

Personal (body) environment

Light boundary

Spaces defined by phenomenological boundaries (e.g., use, thermal, light and other boundaries), which may or may not be coincident or be a physical construction

Intelligent environments: future paradigms

▲ **Figure 8-3** (Continued)

respective actions or responses. Surely a situation of great technical complexity, but with the potential for enormous returns, is if multiple behavior elements that act interactively are considered and implemented. Here the metaphorical neurological model noted above is useful for considering interactive and interdependent multiple behaviors.

8.4 Complex environments

Figure 8–3 summarizes these different past and current paradigms of intelligent environments, and offers a proposal for a future one as well. Which one is right? Which one is the most useful? Under what circumstances would one choose one or the other? These paradigms along with the general discussions above provide a framework for considering more complex environments, although not a model. While many attempts to make 'intelligent spatial environments' focus specifically on one or another approach, the potential richness of combined approaches is clear. The last paradigm shown in Figure 8–3 is intended to express simultaneity and contingency, while relinquishing the idea of a universal system. Our interaction with the multiple environments should be local and discrete, while still maintaining the possibility to slip from one realm to another. It is easy to imagine environments that on the one hand clearly deal with enhancing the physical environment surrounding the human users, while at the same time maintaining approaches that aid in work processes. Interesting questions and opportunities arise when we begin thinking about *interactions* that can occur between the use-centered enhancements and those that deal with the surrounding environment. There is a wealth of understanding available now about how characteristics of surrounding environments affect human activities and tasks. These understandings range from those dealing with basic physiological and psychological responses of humans to different physical environments all the way through specific understandings about how particular kinds of air environments affect humans with certain medical problems.

It is also evident that both levels of cognition and the mode of implementation can vary as well. In this text increasing cognition levels and ever-more embedded or transparent implementation means are signifiers of increasing levels of 'intelligence' in an environment or use environment. The framework provided above gives us a handle on what we might aspire to accomplish within a so-called 'intelligent room'. But, we must not forget that as yet unknown interactions might occur that are not reflected in the framework presented herein. Might we have cognition processes aided by particular kinds of surrounding microenvironments? There are rich possibilities.

Notes and references

1. Cited from the English translation contained in U. Conrad, *Programs and Manifestoes on 20th-Century Architecture* (Cambridge, MA: MIT Press, 1964).
2. Buckminster Fuller, 'The Dymaxion house', *Architecture* (1929), reprinted in J. Krause and C. Lichtenstein, *Your Private Sky: R. Buckminster Fulller* (Lars Mueller Publishers: Baden, 1999).
3. 'The Home of the Near Future (1999), cited from the archives of Koninklijke Philips Electronics NV, located at www.design.philips.com.
4. Colley, M., Clarke, G., Hagras, H, Callaghan, V. and Pounds-Cornish, A. (2001) 'Intelligent inhabited environments: cooperative robotics and buildings', Proceedings 32nd International Symposium on Robotics, Seoul, Korea.

9 Revisiting the design context

In the body of this book, we have examined the very small and the very large. We have also begun to distinguish the different world-views toward smartness and intelligence as practiced by the professions of computer science, materials science, engineering and architecture. Each profession took the micro characteristics of smart materials and addressed them at scales relevant to themselves. While materials scientists went smaller to nano size, engineers and architects have gone much larger, to tangible products and large systems respectively. Essentially, in spite of the radical leap in behavior afforded by smart materials, each profession still understands and applies them through the frameworks that have traditionally defined the use of materials in their field.

The systems framework that typifies the approach the field of architecture has had toward new materials and technologies is somewhat insensitive to innovation and change. In Chapter 7, we noted that even when a new technology has opened the door to unprecedented possibilities, architects and designers often try to make it fit within their normative use. When an advancement in energy technology comes around, we tend to try to use it to improve our HVAC systems; when a new material affords the ability to transiently produce light, we attempt to reconfigure it in the same fashion as our existing light sources.

There is a common belief that technology is deterministic, i.e. that our tools determine our behavior. In the introduction to the book *Living with the Genie: Essays on Technology and the Quest for Human Mastery*, the editors remind us that even though the 'theory of technology would say we devise tools to let us do better what we have to do anyway [. . .] our tools have a way of taking on what seems to be lives of their own, and we quickly end up having to adjust to them.'[1] This view would certainly seem to be supported by the concept of 'technology push' that we described in the first chapter. Yet, oddly enough, the field of architecture is relatively immune to technological domination. This may well have benefits as well as the disadvantages that we have already pointed out about clinging to antiquated technologies long after the underlying science no longer supports them. Architecture is a truly interdisciplinary activity, crossing over many different fields. Besides materials science and engineering, architects must integrate knowledge from all of the sciences with an awareness of cultural developments and history, and balance the requirements of various government agencies, construction practices and economics with concerns for societal responsibility as well as for individual needs. Technology never has, and most likely never will, usurp these multiple roles. Indeed, our built environment might be considered as a bellwether, providing a stable context that allows the freedom of expression and experimentation in so many other parts of our lives.

Designers and architects are in the central position of determining and directing how new developments will enter the world of the everyday. Invariably, as the domain of the built environment is large, extending from buildings to cities to landscapes, we ultimately must think in terms of systems. How, then, should we think about architecture – as the armature for daily life, as the progenitor of tangible artifacts, as the harbinger of fluid environments? The obvious answer is that we must think in terms of all three, but not as a single utopian ideal as we see so often in visions of the future.

Much of this speculation can be rooted in our earlier discussions on boundary. If we recall, the definition of boundary in physical behavior was quite specific: it is the region of energy change between a system and its surroundings. The definition of boundaries between professions, between practices, between the areas of which each has purview is often treated as analogous to the very particular definition of the energy boundary. Either professions are defined as a distinct core of theory with a little fuzziness around the edges where other professions might overlap horizontally, or as a series of hierarchies where each successive practice encompasses a larger and larger area vertically (at, of course, a larger and larger scale of detail). Mechanical engineers and electrical engineers have very distinct theory cores, but overlap where machines and electronics become one and the same. A product designer develops furnishings used in a building designed by an architect built in a city planned by an urban designer in a region studied by a landscape architect. The hierarchic layering tends to be more downward-focused than upward. An architect will be quite aware of the many products used in construction of the building, but relatively unaware of regional issues. In essence, boundaries are drawn for convenience and organization, not for any fundamental characterization of a behavior. While these liberal re-interpretations of the energy boundary might be descriptive of our current modes of professional designation, we cannot turn in reverse and presume that we can use

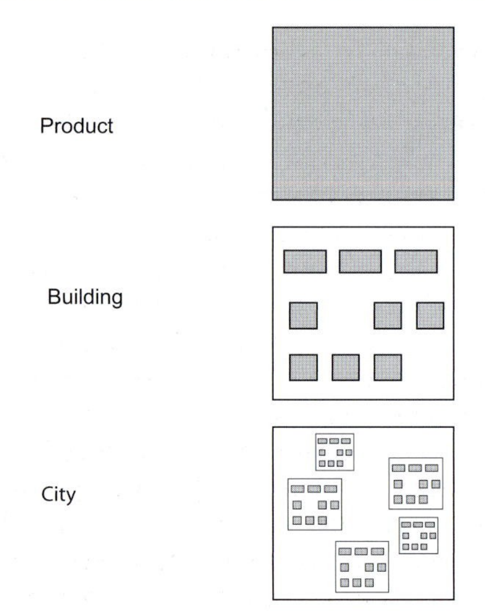

▲ **Figure 9-1** The hierarchical structure that defines the way we do our energy accounting – individual products are added into a building, individual buildings are added to make a city

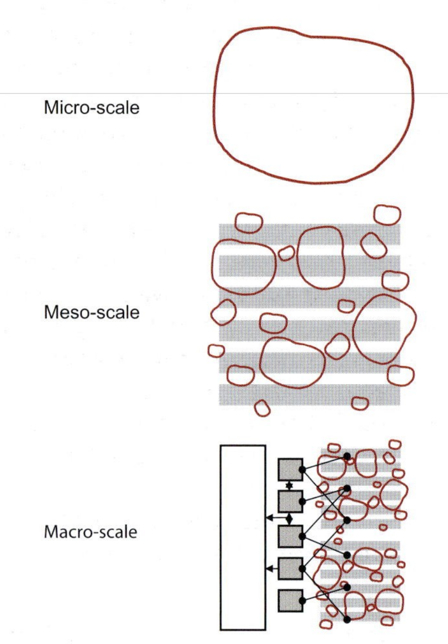

▲ **Figure 9-2** Conceptual relationships between the three major scales of energy systems. Micro-scale is where individual phenomena take place. Meso-scale is the relationship of individual phenomena to different energy forms. Macro-scale are the energy systems responsible for the multiple types of energy

the structure of organizations as an analogy for how an energy system behaves. But this is precisely what we typically do.

A good example of how pervasive our tendencies in this direction have become is the current understanding of energy consumption. No other imperative is more important at this time than the objective to reduce greenhouse gas emissions, and the most effective way to do this is to reduce energy consumption. Many products have been upgraded to reduce their individual energy consumption, and consumers are encouraged to purchase products labeled as efficient. These products, of course, are considered as additive in a building along with each material and system that is part of the construction. *In toto*, then, if each component is 10% more efficient than the average then the presumption is that the building will use 10% less energy. And if several buildings in a region served by a utility are more energy-efficient, then that utility will consume less fossil fuel as a result, and if the different utilities are each using less fuel, then global greenhouse emissions should decrease. Unfortunately, however, when it comes to energy, organizational analogies are not useful. Energy boundaries do not fall into a vertical or horizontal arrangement with a neatly additive accounting system. Buildings, which operate as our primary unit of energy accounting, are not energy systems, nor are they a container of energy systems, of which some are larger, some are smaller, and many straddle the building extents. A building is a unit of private property, nothing more. Recognizing that energy systems layer and network at multiple scales simultaneously, as we see in Figure 9–2, we might begin to imagine the 'building' not as a unit at all, but as a collection of behaviors that intervene at many different locations in the energy network. Nevertheless, we still build, and will continue to do so, in units of buildings, not in units of energy systems.

Does this mean that designers must be fully cognizant of all scales of behaviors and systems, both large and small, in order to operate effectively in their own discipline? We hope not, but we also hope that designers begin to incorporate an understanding of the simultaneity of scales, behaviors, processes and systems as they make their decisions. This may seem to be contradictory, but there is a large difference between appreciating and working with other world-views and system models than there is in having full working knowledge of these other approaches. For example, as we discussed in Chapter 8, many of the visions of the future propose a 'super-environment' in which all aspects are controlled, from the temperature to the sound level, the plasma screen and even the dog. This type of fully contained, fully immersive and fully controlled environment demands that the architect seamlessly integrate products, materials, systems and people at every level. It might only be a building, but the architect is asked to be a master-builder who not only has knowledge of every single component, but mastery as well over each component's production and/or use.

How do we take knowledge, then, from another profession, and apply it to ours? The typical approach is one of extension – we keep expanding our boundaries to overlap with these other fields. But the more we extend, the more we are forced to trade off knowledge for information. This is, of course, what the impetus was for the construction specification system that we discussed in Chapter 2. Knowledge was synthesized, compressed and then basically rewritten as instructions. Education programs have scrambled to keep up with the ever-evolving and growing developments in other professions, and practitioners are signing up for continuing education courses such as 'Construction Law' and 'Mold Remediation'.

Our proposed scenario moves in the other direction, asking designers to relinquish the idea of control over everything in their purview. We would like to trade off a lot of information for some very strategic knowledge. Information is descriptive, and it steadily becomes obsolete as new information arrives. Knowledge, on the other hand, is explanatory. As long as theories stay in place, the fundamental knowledge is elastic enough that an approach to any new information can always be derived. Our intention in the preceding chapters was to lay out a construct in which knowledge served as the explanatory framework and information was simply illustrative of how that knowledge was applied. Energy theory and the basics of material structure are the overarching knowledge fundamentals that govern the approach, while the phenomenological interactions between material behavior and energy environments form the specific knowledge that is representative of the necessary transfer across professions.

The following summarizes some of the key ideas that frame the knowledge you should take away from reading this book:

- Energy is about motion, and motion can only occur if there is a difference in states between a system and its surroundings.
- The exchange of energy can only take place at the boundary between a system and its surroundings.
- Energy must be accounted for during exchange processes. Any energy exchange that is not 100% efficient will

produce heat. As such, all real world processes produce excess heat.

- Usable energy is lost in every exchange. When there is an energy input in one form, the usable energy output is always lower. As an example, when a material absorbs radiation and then releases it, the released radiation will be at a lower energy level (blue wavelength light will degrade to longer wavelength light or infrared).
- Material properties are determined by either molecular structure or microstructure. Any change in a material property, such as what happens in a smart material, can only occur if there is a change in one of the two structures.
- Change can only occur through the exchange of energy, and that energy must act at the scale of structure that determines the material property.

All material behavior can be understood by respecting and adhering to these fundamental principles. For example, most designers have at their fingertips an incredible array of software, helping them visualize their designs, and incorporate more directly the vast amounts of information supplied by other professions. These tools are remarkable, allowing precise visualizations of a given building in terms of its air flow, lighting patterns and structural performance under heavy wind loads. What they do not do, however, is explain

▲ **Figure 9-3** Lighting simulation of interior office space using Radiance software. The contours illustrate the large variability that occurs in typical spaces. Simulation courtesy of John An.

why the results appear as they do. A surprisingly small amount of knowledge regarding materials and material behavior gives the designer enormous power to use the simulation programs for *designing*, rather than evaluating.

The fundamentals are what we would wish you, the reader, to know. But what do we want you to think about and to do? This book was intended to present a roadmap for how we, as designers and architects, might begin selectively to appropriate knowledge from other fields and use it to ask new questions about our own. By providing a clear ground plan in knowledge, and overlaying it with applications, we have tried to present an open-ended map that enables the designer to make more of his or her own selections, combinations, products and/or systems, rather than providing a prescriptive set of directions simply to instruct in the implementation of these new materials and technologies. Armed with this knowledge, a designer should then be able to use, and develop, any assembly of any component that has a dynamic behavior.

This knowledge, of course, is not exclusive to smart materials and new technologies. Almost every type of behavior that we create through the manipulation of physical phenomena can be reproduced with more conventional methods and materials. As an example, consider the Aegis Hyposurface Installation by dECOi Architects. The intent of this 'wall' was the translation between the intangible information world and the very tangible, and omnipresent, tactile environment. Described as 'a giant sketch pad for a new age', this large surface becomes everything that the typical architectural surface is not – it is interactive and responsive in real time to physical stimuli, including sound, touch, light and motion. Pixels the size of one's hand, activated by a large network of conventional pneumatics, become the tangible media that register change through their relative motions. Although more of a didactic piece than a functional work, the Hyposurface nevertheless is an excellent example of how the concepts of immediacy, transiency, self-actuation, selectivity and directness can supersede the available technologies. Smart materials would certainly give us a leg up on this sort of design – they would be seamless, discrete, efficient – but our point here is that thinking and designing in response to actions and behaviors instead of in terms of artifacts and things is facilitated by, rather than restricted to, advancements in materials and technologies.

By focusing on phenomena and not on the material artifact we may be able to step out of the technological cycle of obsolescence and evolution. This is particularly important as,

▲ **Figure 9-4** Aegis Hyposurface by deCOi Architects. Each moving element of the panel was driven by pneumatic actuators

given the long lifetimes of buildings, we must be evermore nimble to avoid cementing in place an obsolete technology. As a result, this approach requires a much more active engagement by the reader than does the typical technology textbook or materials compendium. Indeed, we fully recognize that these materials, products and systems will quickly become obsolete. Our intention is for you to understand them in relation to the phenomena and environments they create. The electrochromic glazing currently being proposed for buildings will almost assuredly be phased out, or at least substantially altered from its current version. If knowledge of a material or system is tied only into an account of its specifications and a description of its current application, then that knowledge becomes obsolete along with the material. By operating at the level of behavior and phenomena, we will recognize that a particular technology at any given time is only illustrative of the possibilities, not their determinant. As the materials and products cycle through evolution and obsolescence, the questions that are raised by their uses should remain.

As an example of these cycles, the term 'nanotechnology' has rapidly become commonplace over the past few years, steadily supplanting the ubiquitous and soon to be overused term of 'smart' technologies. This shift in focus will be an interesting test for the utility and longevity of this book's contents. In Chapter 3 we noted that the behavioral type for thermal phenomena switches from continuum to non-continuum at the micron scale and smaller. All of our discussions about materials and investigations of material properties thus far have been couched in continuum behavior. We may have thought that the 'First Principles' in continuum mechanics were complex enough – indeed we were not able even to apply the Navier–Stokes equations practically until a few decades ago – but non-continuum mechanics, i.e. quantum mechanics, is another beast altogether, requiring a very different level of knowledge fundamentals for operation in the material world. It would be overstepping the mark to suggest that architects and designers should acquire this additional knowledge. Nevertheless, many of the questions currently raised by nanotechnology may fit into the framework that we have been developing to 're-think' the design professions' approach to materials. Consider the following comments made by Paulo Ferreira, a materials science researcher at the University of Texas:

> What is so special about nanotechnology? First of all, it is an incredibly broad interdisciplinary field. It requires expertise in physics, chemistry, materials science, biology, mechanical and electrical engineering, medicine and their collective knowledge. Second, it is the boundary between the atoms and the molecules, and the macroworld, where ultimately the properties are dictated by the fundamental behavior of atoms. Third, it is one of the final great challenges for humans, for which control of materials at the atomic level is possible.[2]

He underscores four of the key aspects of the approach we developed throughout the course of this book – the multi-disciplinary exchange of knowledge, the exploration of the relationships between multiple scales and their differing behaviors, the understanding that material properties are dictated at the smallest scale, and thus recognition that the overarching macro-scale behavior can be controlled by underlying nano-scale design. This last aspect might be not only the most provocative, but also the most indicative of the design impact of smaller-scale technologies as compared to our more normative, and more visibly present, technologies. The concept of bottom-up that was briefly discussed in Chapter 2 requires that one begins at the smallest point and then builds up from there. Rather than being chosen after the basic design is completed, materials and properties become the starting point. So how does one even begin to think about the impact of nanotechnology when you can make anything with any behavior?

While the basic physics and applications for nanotechnology might be radically different from those of smart materials, the method for framing and asking questions may well be the same. The questions we should be thinking about include determining the root need or the underlying problem. As designers, we often limit our problem definition to possible actions within our domain. We specify the glazing that allows for maximum daylight and minimum heat gain, presuming that these needs are *prima facie*. Engineers try to ameliorate the thermal and lighting problem caused by a glazed façade with additional systems because they assume that this is a requirement of building design. The material scientist searches for a new material that combines the transparency needed for light with the insulating value needed for heat because it is the best solution to optimize the glazed façade. Who steps back to ask why? For what end? Interestingly enough, when each profession constrains its problems to those within its own domain, professionals fall prey to assuming that the activities and problems being addressed outside of their domain are inviolate, not open to question. The question being wrestled with in one profession becomes fact to another.

While we often think of design as a creative profession with few constraints, the reality is that most of the design professions are regulated. Sometimes it is the process that is regulated, such as in architecture and landscape architecture, and sometimes it is the product that is regulated, as in industrial and product design. While few designers would be content to operate solely within the narrow bounds determined by law, their responsibilities nevertheless stop at the border of those bounds. The design professions are thus in an interesting predicament, for it is only through the process and products of design that the advancements in other professions can be made physically manifest in the human world. This manifestation introduces another level of responsibility, for it means that each profession must think beyond the extents of their production. A building cannot be treated as an autonomous object; the architect must also think about its impact and interaction with a variety of systems that no one would consider remotely architectural. So while the profession's knowledge might be defined and confined, its implications touch most aspects and scales of the human environment.

Our foray into the promise of smart materials and new technologies was more than a technical recounting of properties and products, it was also an attempt at a transactional language that would enable us to ask questions from within our own profession that would directly impact or involve many others. What if, instead of selecting between various materials and technologies that come to us from without, we could articulate a problem from within that would engage other professions? Rather than simply choosing between glazing materials that transmit light, we might be able to articulate a more fundamental problem that would call into question the use of glazing altogether.

If we think back to those characteristics that we identified in the first chapter – immediacy, transiency, self-actuation, selectivity and directness – we would probably begin to recognize that the search for a comprehensive model to describe the intelligent environment is not only foolhardy, it neglects the most important interaction of all – that of humans with each other. Regardless of which type of activity a designer practices, the products that each produces have no value or relevance until they enter the public domain. Our results, then, are not buildings or urban infrastructure, but places of human interaction, and as such, always subject to public reinterpretation of our intentions.

Several years ago, one of the co-authors of this book was teaching design at an architecture school in Philadelphia. The students, who were at an advanced level, received a real commission – the design of a new community center in North Philadelphia. Before meeting with community leaders, the students developed several design concepts that they hoped to discuss at the meeting. Features that we agreed were important to highlight included the establishment of a neighborhood identity, a mediation of scale between the domestic and the public, and creation of an entryway that encouraged inclusivity of all the diverse residents. At the meeting, the community leaders told us what they thought was most important: the ability to keep the building clean with little effort (they asked for tiled surfaces throughout so that it could be easily washed), and the need for security at all times.

How do we reconcile our view of a future populated with remarkable materials and configured for seamless communication with the reality of the human condition? In *Living with the Genie*, Daniel Sarewitz and Edward Woodhouse suggested that elaborate visions of the future were the province of the wealthy few who could indulge in such speculation: 'Thus far, the exuberant vision to remake the world with nanotechnology has come from committees drawn from a small group of experts, mostly male, mostly upper middle class, universally in possession of great technical expertise.' Smartness, rather than having a clear definition, may well be in the eye of the beholder.[3]

Certainly, many of the materials and products that we have explored in this book are economically beyond the reach of the majority of building and infrastructure projects, and many as well can only be described as frivolous. The conditions and implications, however, of these materials can reach through to every design act, at every level. The quest for selectivity, directness, immediacy, transiency and self-actuation might actually allow the expansion of the design realm more widely into the greater public domain. If, by applying the fundamentals, we can reduce energy use by a factor of ten by discretely acting only where necessary, then all will benefit. If we can move away from the overarching idea of a fully interconnected, and thus controlled, infrastructure, and operate discretely and locally, then many of the advantages offered by new technologies can be appropriated by a greater diversity of projects.

The potential, however, for rethinking our normative deployment of materials extends far beyond the notions of efficiency and expediency. In Chapter 1, we suggested that the advent of smart materials would eventually enable the design of direct and discrete environments for the body. What does this mean in the context of the chapters that followed? Fundamentally, it means that design begins with a single, small action. Rather than designing the static shell of the building, and then progressively moving smaller, with each step in the process geared toward greater delineation of the design artifacts, we may have the opportunity to move in the opposite direction. We now have technologies that can do anything, even though they would rarely be visible. The artifact could support the design intent, instead of being its physical manifestation. We come back to the questions of what the experience could be, what the occupants should feel, how they would interact with their surroundings. Instead of designing at a large scale to produce ancillary effects, we might be able to design at the small scale to produce a larger human experience.

When we first began teaching courses in smart materials, we derived an expression for what we considered to be their ultimate goal:

direct and immediate action at the precise location so desired.

We still think so today.

Notes and references

1. Lightman, A., Sarewitz, D. and Dresser, C. (2003) 'Introduction', in *Living with the Genie: Essays on Technology and the Quest for Human Mastery*. Washington, DC: Island Press, pp. 1–2.
2. Ferreira, Paulo J. (2004) 'Nanomaterials', in J. Brito, M. Heiter and R. Rollo (eds), *Engineering in Portugal during the 21st Century*. Lisbon: Don Quixote, p. 3.
3. Lightman *et al.*, *Living with the Genie*, p. 67.

Glossary

Absorptance (acoustic)	the dimensionless ratio of incident vibrational energy that has been converted to another energy form, such as heat, to the total incident energy on a material surface. The working definition of absorptance is slightly different: the dimensionless ratio of incident vibrational energy that is *not reflected* to the total incident energy on the surface. A perfect absorber with a reflectance of 1 reflects no energy – all the incident energy may be absorbed or transmitted.
Absorptance (luminous)	the dimensionless ratio of incident radiant energy (in the visible spectrum) that has been converted to another energy form, such as heat, to the total incident energy on a material surface. A perfect absorber with a reflectance of 1 reflects and transmits no light.
Actuator	a control element that is driven by a signal, often electrical, that produces enough power to operate a mechanical element, such as a valve. Common actuator types are electromechanical, hydraulic and pneumatic.
Aerogel	generically describes any colloidal solution of a gas phase and solid phase. More typically, aerogel refers to a specific material.
Artificial Intelligence	programs that can perform activities that are typically associated with human intelligence, such as recognition.
Augmented reality	a composite view constructed of a real scene overlaid or augmented with a virtual scene.
Biomimetic	the imitation of nature or the study of the structure and function of biological substances.
Birefringence	Occurs when an anisotropic material possesses different refractive indices depending on how the incident light is polarized.
Bioluminescence	light produced by living organisms through an enzymatic chemical reaction.
Biosensor	a general designation that refers to either a sensor to detect a biological substance or a sensor that incorporates the use of biological substances in its construction.
Chemochromics	materials that change their color in response to changes in the chemical composition of their surrounding environment.

Cladding	the outer sheathing of a building that provides the final layer of the envelope. The cladding is exposed to weather and thus needs to be durable while, simultaneously, it is the cladding that is most responsible for a building's appearance.
Composite	a multi-component material produced when metal, ceramic or plastic materials provide a macrostructural matrix for the distribution of strengthening agents, such as filaments or flakes, throughout the material, increasing its structural performance. Each component, however, maintains its properties.
Conduction (electrical)	the transmission of electricity through the movement of electrons.
Conduction (thermal)	the diffusive transfer of heat and mass, through direct molecular contact.
Conductive polymers	organic materials that conduct electricity.
Convection	specific motion in a fluid material that results in heat and mass transfer.
Copolymer	a polymer that consists of two or more distinct monomer units that are combined along its molecular chains, in block, graft or random form.
Critical angle	the smallest angle of incidence that will produce total internal reflection at an interface boundary between two mediums with different refractive indices.
Curtain wall	an exterior non-load bearing skin of a building.
Detector	a device that responds to a change in some energy – usually light – and produces a readable signal.
Dichroism	a diochroic material that has selective spectral absorption that differentiates its transmissive spectrum from its reflective spectrum.
Dielectric	a material that is electrically insulating, i.e. a very weak conductor.
Distributed intelligence	the distribution of intelligent entities throughout a system, with no distinct center.
Doping	the addition of donor or acceptor impurities into a semiconductor material to increase its conductivity.
Elastomers	polymers that have largely amorphous structures, but are lightly cross-linked, and are thus able to undergo large and reversible elastic deformations.
Electrochromics	materials that change their color in response to changes in an electric field; often used to change the transparency of glass laminates.
Electroluminescents	materials that luminescence or emit light when subjected to an electric field.

Electromagnetic radiation a large family of wave-like energy that is propagated at the speed of light. The electromagnetic spectrum encompasses wavelengths from as small as gamma rays to as large as radio waves.

Electrostriction the change in shape produced when a dielectric material undergoes strain when subjected to an electrical field.

Electrorheological ER fluids contain micron-sized dielectric particles in suspension. When exposed to an electrical field, an ER fluid undergoes reversible changes in its rheological properties including viscosity, plasticity and elasticity.

Emergent intelligence an intelligent system that is bottom up, emerging from simpler systems.

Emissivity the measure of the ability of a surface to emit thermal radiation relative to that which would be emitted by an ideal 'black body' at the same emperature.

Envelope the term describes the three-dimensional extents of a building.

Extrinsic property a material property that depends on the amount or conditions of material present. Whereas density is intrinsic, mass is extrinsic.

Ferroelectricity the alignment of electric dipoles in a material to produce spontaneous polarization when it is subjected to an electric field.

Ferromagnetism the alignment of magnetic dipoles in a material to produce spontaneous polarization when it is subjected to a magnetic field.

Fiber-optics strands, cables or rods that carry light by internal reflection; used in lighting and communications. The fibers can be glass or of PMMA.

Fluorescence fluorescence is the property of some atoms and molecules to absorb light at a particular wavelength (higher energy) and to emit light (luminescence) of longer wavelength. If the luminescence disappears rapidly after the exciting source is removed, then it is termed fluorescence, but if it persists for a second or more, it is termed phosphorescence.

FOLED flexible organic light-emitting devices built on flexible substrates typically used for flat panel displays.

Fresnel lens a type of flat lens with a concentric series of simple lens sections that either focus parallel light rays on a particular focal point or, alternatively, generate parallel rays from a point source.

Gels any semi-solid system in which liquid is held in a network of solid aggregates.

Haptics the production of a tactile sensation, such as heat and pressure, at the interface between a human and a computer.

Health monitoring (structural) the comparison of the current condition to earlier conditions to proactively predict potential failure. Most often used for large structures such as bridges and building foundations.

HVAC an acronym for heating, ventilation and air conditioning.

Hydrogels three-dimensional molecular structures that absorb water and undergo large volumetric expansion.

Illuminance the density of light flux on a surface, the ratio of incident flux to the area of the surface being illuminated.

Incandescence the production of light through heat.

Index of refraction the ratio of the velocity of light in a vacuum to the velocity of light in a particular medium.

Inorganic defined as any compound that is not organic.

Intelligent agent software that can perform tasks without supervision.

Internal reflection the process through which light travels within a high refractive index medium.

Intrinsic property a material property that is independent of the quantity or conditions of the material.

Inverse Square Law applies to all radiant propagation from a point source, including that produced by sound and light. The intensity diminishes with the square of the distance traveled.

Laser an acronym for light amplification by the stimulated emission of radiation. A quantum device for producing coherent (parallel) light.

LCD liquid crystal display. The typical display sandwiches a liquid crystal solution between two polarizing sheets. When electric current is applied to the crystals, they are aligned in such a manner so as to block transmitting light.

LED light-emitting diode. A semiconductor device that releases light during the recombination process.

Light pipe although occasionally used to refer to light guides or fiber-optics, the primary use of the term in buildings is for a hollow macro-scaled device that transports light through reflection and refraction.

Liquid crystals anisotropic molecules that tend to be elongated in shape and that have an orientational order that can be changed with the application of energy.

Luminance the light flux that is reflected from a surface.

Luminescence the emission of light from a substance when electrons return to their original energy levels after excitation. Luminescence is an overarching term referring to any light production that involves the release of photons from electron excitation.

Luminescents materials that emit non-incandescent light as a result of a chemical action or input of external energy.

Magnetorheological MR fluids go from fluid to solid when subjected to a magnetic field due to a change in their rheological properties, including viscosity, plasticity, and elasticity.

Magnetostrictive materials that change dimension when subjected to a magnetic field or that generate a magnetic field when deformed.

Mechatronic a term generically used to describe electronically controlled mechanical devices (mechanical-electronic).

MEMS microelectronic machines; typically small devices based on silicon chip technologies that combine sensing, actuating and computing functions. The term is an acronym for micro-electromechanical system but today almost any micro-scaled device is referred to as a MEMS device.

MesoOptics™ a type of coating or film with holographically generated microstructural diffusers that produce optical control of the transmitting light.

Meso-scale length dimensions on the mm to cm scale. Often referred to as miniature.

Microencapsulation individually encapsulated small particles or substances to enable suspension in another compound.

Micromachine a structure or mechanical device with micro-scale features.

Microprocessor the IC-driven arithmetic logic of a computer.

Micro-scale length dimensions on the micrometer to 0.1 mm scale.

Microstructure the structural features of a material such as its grain boundaries, its amorphous phases, grain size and structure.

MOEMS micro-electro-optical mechanical systems; MEMS with optics.

Nanotechnology the exploitation of the property differences between the scale of single atoms to the scale of bulk behavior. Also, the fabrication of structures with molecular precision.

NEMS nanoscale MEMS at scales of 1000 nm or less.

Nitinol a nickel–titanium alloy used as a shape memory alloy.

OLED organic light-emitting devices made from carbon-based molecules rather than from semiconductors.

Optoelectronics the combination of optical elements, such as lasers, with microelectronic circuits.

Organic a term applied to any chemical compound containing carbon as well as to a few simple carbon-based compounds such as carbon dioxide.

Pervasive computing	when computational and interactive devices are seamlessly integrated into daily life.
Phase change	the transformation from one state (solid, liquid, gas) to another.
Phase transformation	change that occurs within a metal system, most often refers to a change in crystalline structure.
Phosphorescence	luminescence that remains for more than a second after an electron excitation.
Photochromics	materials that change their color in response to an energy exchange with light or ultraviolet radiation.
Photodiode	semiconductor diode that produces voltage (current) in response to a change in light levels.
Photoelectrics	devices based on semiconductor technologies that convert light (radiant) energy into an electrical current.
Photoluminescence	the luminescence released from a material that has been stimulated by UV radiation.
Photoresistors	devices based on semiconductor technologies in which the absorption of photons causes a change in electrical resistance.
Photovoltaic effect	the production of voltage across the junction of a semiconductor due to the absorption of photons.
Piezoceramic	ceramic materials that possess piezoelectric properties.
Piezoelectric effect	the ability of a material to convert mechanical energy (e.g., deformation induced by a force) into electrical energy and vice-versa.
Polarization	occurs when the centers of the positive and negative charges are displaced, thereby producing an electric dipole moment.
Polarized light	electromagnetic radiation, primarily light, in which the wave is confined to one plane.
Privacy film	a type of film that is transparent from particular view angles and opaque from other angles (often called *view directional film*).
Pyroelectric materials	materials in which an input of thermal energy produces an electrical current.
Radiant energy	electromagnetic energy as photons or waves.
Radiation	the emission of radiant energy.
Reflectance	the ratio of reflected to incident radiation.
Reflection	the amount of light leaving a surface. Surfaces are subtractive, so the amount of reflected light must always be less than the arriving or incident light. Furthermore, the angle of incidence is equal to the angle of reflection.

Refraction the bending of a light wave when it crosses a boundary between two transparent mediums with different refractive indices.

Reverberation reverberation is the continuance of collected sound reflection in a space. The reverberation time is the amount of time it takes for a sound level to drop by 60 dB after it has been cut off.

Self-assembly self-assembly (also called Brownian assembly) results from the random motion of molecules and their affinity for each other. It also refers to bottom up molecular construction.

Semiconductor a nonmetallic material, such as silicon or germanium, whose electrical conductivity is in between that of metals and insulators, but it can be changed by doping.

Sensor a device that quantifies its energy exchange to provide measurement of an external energy field.

Shape memory effect the ability of a material to be deformed from one shape to another and then to return to its original shape after a change in its surrounding stimulus environment (e.g., thermal, magnetic). In metals, this phenomenon is enabled by a phase transformation.

Shape memory alloys metal alloys, e.g., nickel–titanium, that exhibit the shape memory effect.

Shape memory polymers polymeric materials that exhibit the shape memory effect.

Snell's Law the relationship between angles of incidence and refraction between two dissimilar mediums.

Spectral absorptivity wavelength-specific absorption. Reflectivity and transmissivity are often wavelength-specific as well. Most materials have uneven absorption spectra.

Suspended particle display or SPD, a suspension of randomly oriented particles that can be oriented under application of a current.

Thermochromics materials that change their color in response to a thermal energy exchange with the surrounding thermal environment.

Thermoelectric effect the conversion of a thermal differential into a current (Seebeck effect) and vice versa (Peltier effect).

Thermophotovoltaic a device that converts longwave thermal radiation into electricity.

Thermotropics materials that change their optical properties due to a thermally produced phase change.

Thin films a large class that is commonly used to refer to any thin amorphous film of semiconductor layers.

Total internal reflection a phenomenon that occurs at the interface between two mediums when light at a small angle (below the critical angle) is passing from a slow medium to a fast medium.

Transducer the conversion of the measured signal into another, more easily accessible or usable form.

View directional film a type of film that is transparent from particular view angles and opaque from other angles (often called *privacy film*).

Wavelength the distance traveled in one cycle by an oscillating energy field propagating in a radiant manner. The peak to peak distance between one wavecrest and the next.

Bibliography

Addington, D.M. (1997) Boundary Layer Control of Heat Transfer in Buildings. Doctoral thesis, Harvard University.

Addington, D.M. (2001) The history and future of ventilation. In Spengler, J., Samet, J. and McCarthy, J. (eds), *Handbook of Indoor Air Quality.* New York: McGraw–Hill.

Addington, M., Kienzl, N. and Schodek, D. (2002) *Smart Materials and Technologies in Architecture.* Harvard Design School Design and Technology Report Series 2002-2. Cambridge, MA.

Antonelli, P. (1995) *Mutant Materials in Contemporary Design.* The Museum of Modern Art. New York: Department of Publications, the Museum of Modern Art.

Ashby, M.F. (2002) *Materials and Design: The Art and Science of Material Selection.* Oxford: Butterworth–Heinemann.

Banham, R. (1984) *The Architecture of the Well-Tempered Environment,* 2nd edn. Chicago: University of Chicago Press.

Banks, J. (ed.) (1998) *Handbook of Simulation.* New York: John Wiley & Sons.

Barrett, C.R., Nix, W.D. and Tetlman, A.S. (1973) *The Principles of Engineering Materials.* Englewood Cliffs, NJ: Prentice-Hall.

Braddock, S.E. and Mahoney, M.O. (1998) *Technotextiles.* London: Thames and Hudson.

Brosterman, N. (1999) *Out of Time: Designs for the Twentieth-Century Future.* Harry N. Abrams.

Button, D. and Pye, B. (1993) *Glass in Building: A Guide to Modern Architectural Glass Performance.* Oxford: Butterworth Architecture.

DePree, C.G. and Axelrod, A. (eds) (2003) *Van Nostrand's Concise Encyclopedia of Science.* New York: John Wiley and Sons.

De Valois, R.L. and De Valois, K.K. (1990) *Spatial Vision.* Oxford: Oxford University Press.

Dyson, F. (1997) *Imagined Worlds.* Boston, MA: Harvard University Press.

Feynman, R.P., Leighton, R.B. and Sands, M. (1963) *The Feynman Lectures on Physics*, vols I, II, and III. Reading, MA: Addison–Wesley.

Fitch, J.M. (1972) *American Building 2: The Environmental Forces That Shape It.* Boston, MA: Houghton–Mifflin.

Gersil, K.N. (1999) *Fiber Optics in Architectural Lighting.* New York: McGraw–Hill.

Guy, A.G. (1976) *Essentials of Materials Science.* New York: McGraw–Hill.

Halliday, D. (1997) *Fundamentals of Physics,* 5th edn, vols I and II. New York: Wiley.

Hewitt, G.F., Shires, G.L. and Polezshaev, Y.V. (eds) (1997) *International Encyclopedia of Heat and Mass Transfer.* Boca Raton, FL: CRC Press.

Hummell, R.E. (1998) *Understanding Materials Science: History, Properties, Applications.* New York: Springer.

International Energy Agency (2000) *Daylight in Buildings.* LBNL-47493. Berkeley, CA: IEA.

Jandl, H.W. (1991) *Yesterday's Houses of Tomorrow.* Washington, DC: The Preservation Press.

Khartchenko, N.V. (1998) *Advanced Energy Systems.* London: Taylor & Francis.

Kiemzl, N. (2002) Evaluating Dynamic Building Materials. Doctoral thesis: Harvard University.

Kutz, M. (1998) *Mechanical Engineer's Handbook.* New York: John Wiley & Sons.

Lightman, A., Sarewitz, D. and Desser, C. (2003) *Living With the Genie: Essays on Technology and the Quest for Human Mastery.* Washington, DC: Island Press.

Livingstone, M. (2002) *Vision and Art: The Biology of Seeing.* New York: Harry N. Abrams.

Lupton, E. (2002) *Skin: Surface, Substance and Design.* Princeton, NJ: Architectural Press.

McCarty, C. and McQuaid, M. (1998) *Structure and Surface: Contemporary Japanese Textiles.* New York: Department of Publications, the Museum of Modern Art.

Messenger, R. and Ventre, J. (2000) *Photovoltaic Systems Engineering.* Boca Raton, FL: CRC Press.

Mitchell, W.J. (2003) *ME++ The Cyborg Self and the Networked City.* Cambridge, MA: The MIT Press.

Mori, T. (ed.) (2002) *Immaterial/Ultramaterial: Architecture, Design and Materials.* New York: George Braziller.

Palmer, S.E. (2002) *Vision Science: Photons to Phenomenology.* Cambridge, MA: MIT Press.

Schodek, D., Bechthold, M., Griggs, K., Kao, K., and Steinberg, M. (2004) *Digital Design and Manufacturing: CAD/CAM Applications in Architecture.* New York: John Wiley & Sons.

Schodek, D.L. (2001) *Structures,* 4th edn. Englewood Cliffs, NJ: Prentice Hall.

Schwartz, M. (ed.) (2002) *The Encyclopedia of Smart Materials,* vols I and II. New York: John Wiley and Sons.

Smith, W. (1986) *Principles of Materials Science and Engineering.* New York: McGraw–Hill.

Tritton, D.J. (1988) *Physical Fluid Dynamics.* Oxford: Oxford University Press.

Van Wylen, G.J. (1994) *Fundamentals of Classical Thermodynamics,* 4th edn. New York: Wiley.

Index